Practical

Springer
London
Berlin
Heidelberg
New York
Barcelona
Budapest
Hong Kong
Milan
Paris
Santa Clara
Singapore
Tokyo

Other titles in this series

The Observational Amateur Astronomer
Patrick Moore (Ed.)

The Modern Amateur Astronomer
Patrick Moore (Ed.)

Telescopes and Techniques
C.R. Kitchin

Small Astronomical Observatories
Patrick Moore (Ed.)

The Art and Science of CCD Astronomy
David Ratledge (Ed.)

The Observer's Year
Patrick Moore

Seeing Stars
Chris Kitchin and Robert W. Forrest

Photo-guide to the Constellations
Chris Kitchin

The Sun in Eclipse

Michael Maunder and
Patrick Moore

Springer

Cover illustrations: Front cover: A montage capturing the "miracle" of the Sun in eclipse – a Pinatubo volcanic ash sunset seen in Namibia, together with a solar eclipse seen in Kenya in 1980 *(Michael Maunder)*. Back cover (background): A solar eclipse seen in Rajasthan in 1995 *(Peter Cattermole)*. Back cover (inset): The "Diamond Ring" effect seen at the 1976 eclipse in Zanzibar *(Michael Maunder)*.

ISBN 3-540-76146-2 Springer-Verlag Berlin Heidelberg New York

British Library Cataloguing in Publication Data
Maunder, Michael
    The sun in eclipse. – (Practical astronomy)
    1. Solar eclipses
    I. Title   II. Moore, Patrick, 1923–
    523.7'8
ISBN 3540761462

Library of Congress Cataloging-in-Publication Data
A catalog record for this book is available from the Library of Congress

Typeset by EXPO Holdings, Malaysia
Printed and bound at the University Press, Cambridge
58/3830–54321   Printed on acid-free paper

Foreword

Total eclipses of the Sun are always fascinating. They no longer cause alarm in civilised countries, but they do excite a tremendous amount of interest. They are also scientifically important, since they allow Earth-based observers to view phenomena which can never be properly seen at other times.

As seen from any one location on the Earth's surface, total solar eclipses are rare; thus England saw no totality between 1927 and 1999, and then must wait until 2090. Therefore, would-be eclipse viewers must travel to other parts of the world. This means careful preparation.

In this book, we have done our best to explain what eclipses are; what phenomena can be seen; what plans must be made; and – most importantly – what are the procedures for scientific work and photography. Full instructions are given based on the authors' personal experience, and we hope that what we have written will be of use to others.

This is an appropriate time to produce a book of this nature, because on 11 August 1999 a total eclipse will be seen from parts of Cornwall and Devon, as well as Alderney in the Channel Islands. So let us hope for clear skies – and if we are unlucky, prepare to journey to other countries for the eclipses due in the next few years.

Michael Maunder
Patrick Moore

Contents

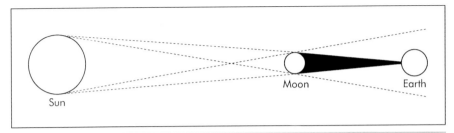

Total eclipse The umbral shadow of the Moon reaches all the way to the Earth. Outside the umbra, approximately one third of the Earth's daylit surface will see a partial eclipse.

How it looks

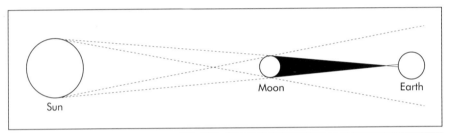

Annular eclipse In this type of eclipse, the umbral shadow falls short of the Earth's surface, because the Moon is too far away to obscure the Sun entirely.

How it looks

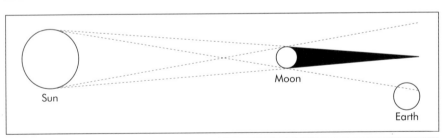

Partial eclipse In this type of eclipse, the Moon's umbra misses the Earth entirely. Viewed from the Earth, all that is visible is the Sun partially covered by the Moon.

How it looks

Chapter 1

Sun, Moon and Stars

Introduction

On 11 August 1999 there will be a total eclipse of the Sun. The track of totality will cross England, and – clouds permitting! – people in parts of Cornwall and Devon will be able to watch the first English total eclipse since 1927. They will see the dark disk of the Moon encroach upon the brilliance of the Sun, finally blotting it out and revealing the glory of the solar atmosphere, with the pearly corona and the red prominences which look so uncannily like flames. The sky will darken, and for a few fleeting moments it will almost seem as though Nature has come to a halt.

Of course eclipses have been observed since very ancient times, and to our ancestors they were very alarming phenomena. Even today there are still some people who are by no means sure how eclipse happens. A basic understanding of the make-up of the universe must be the first step in the explanation, so let us begin with a brief description of the main characters in the cosmic story.

The Earth upon which we live is a planet, 7926 miles (12,756 km) in diameter, moving round the Sun at a distance of 93,000,000 miles (150,000,000 km). The Sun itself is an ordinary star, and all the stars we can see on any clear night are themselves suns, many of them far larger, hotter and more luminous than ours. They are, of course, much further away, and their remoteness means that ordinary units of measurement such as the mile and the kilometre become hopelessly inconvenient

– just as it would be cumbersome to give the distance between London and Manchester in inches. Instead, we use the light-year. Light does not travel instantaneously; it flashes along at 186,000 miles (300,000 km) per second, so that in a year it can cover almost 6 million million miles (9.4 million million km). This is the astronomical light-year, which please note, is a measure of distance, not of time. Even the nearest star beyond the Sun is over 4 light-years away. On the other hand it takes light only 8.6 minutes to reach us from the Sun, and it can pass between the Earth and the Moon in less than two seconds.

The Planets

The Sun is the centre of the Solar System – our home in space. There are nine planets, of which the Earth comes third in order of distance. The planets have no light of their own, and shine only because they are being lit up by the Sun, so that if some malevolent demon suddenly snatched the Sun out of the sky the planets would vanish too – though the stars would be completely unaffected.

Any rough plan of the Solar System shows that it is divided into two well-marked parts. First there are four relatively small, rocky planets: Mercury, Venus, the Earth and Mars. Next comes a region in which move many thousands of dwarf worlds known as minor planets or asteroids, and then follow four giants: Jupiter, Saturn, Uranus and Neptune, together with a peculiar little body, Pluto, which may not be worthy of true planetary status. The main details about the planets are best summarised in Table 1.1.

Most planetary orbits are not very different from circles, but Pluto is an exception; its distance from the Sun ranges between 2766 million and 4583 million miles (4425 and 7375 million km), so that it spends part of its time closer-in than Neptune; between 1979 and 1999 Neptune, not Pluto, is the outermost member of the planetary family. Pluto is also exceptional in another way. In general the planets move roughly in the same plane, so that if you draw a plan of the Solar System on a flat table you are not very far wrong, but Pluto's orbital inclination is 17 degrees, so that there is no fear of a collision with Neptune.

The first five planets have been known since early times, because they are naked-eye objects. Only Mercury

Table 1.1. The planets of our Solar System

Planet	Mean Distance from the Sun millions of		Orbital Period	Axial Rotation (equatorial)	Equatorial Diameter		$\dfrac{Mass_{Planet}}{Mass_{Earth}}$
	miles	km			miles	km	
Mercury	36.0	57.9	88 days	58.2 days	3030	4878	0.055
Venus	67.2	108.2	224.7 days	243.2 days	7523	12,104	0.816
Earth	93.0	148.6	365.3 days	23h 56m 4s	7926	12,756	1
Mars	141.5	227.9	587.0 days	24h 37m 22s	4222	6794	0.107
Jupiter	483	778	11.9 years	9h 50m 30s	89,424	143,884	1319
Saturn	886	1427	29.5 years	10h 13m 59s	74,914	120,536	744
Uranus	1783	2870	84.0 years	17h 14m	31,770	51,118	14.6
Neptune	2793	4497	164.8 years	16h 6m	31,410	50,538	17.2
Pluto	3666	5900	247.7 years	6d 9h 17s	1444	2324	0.002

is not obtrusive, because it always keeps in the same part of the sky as the Sun and can never be seen against a really dark background; Venus and Jupiter far outshine any star, and so does Mars at its best, while Saturn is bright enough to be very conspicuous. The three outermost planets have been discovered only in modern or near-modern times: Uranus in 1781, Neptune in 1846 and Pluto as recently as 1930. Uranus can just be seen with the naked eye if you know where to look for it, and binoculars will show Neptune, but Pluto is very much fainter.

The stars are so far away that their individual movements are very slight; the constellation patterns which we see today are to all intents and purposes the same as those which must have been seen by Julius Cæsar, Homer or even the builders of the Pyramids. The planets, however, wander around from one constellation to another; because they are moving roughly in the same plane (again with the exception of Pluto) they keep to a well-defined band round the sky known as the Zodiac.

Each planet has its own special characteristics. Mercury is rocky and barren, with almost no atmosphere; Venus has a very dense atmosphere, made up chiefly of the gas carbon dioxide – and since the clouds contain large quantities of sulphuric acid, any form of life there is most improbable! Telescopically, both Mercury and Venus show phases, or apparent changes of shape from new to full. Obviously the Sun can illuminate only half of a planet at any one time, and our view depends upon how much of the sunlit side is turned in our direction. Very occasionally these two planets can pass in transit across the face of the Sun, and appear as black disks; Mercury will transit on 15 November 1999, Venus not until 8 June 2004.

Mars is distinguished by the strong red colour which led the ancients to name it in honour of the God of War. Its poles are covered with white ice-caps, and there are dark areas, where the reddish dusty material has been blown away by winds in the very thin Martian atmosphere. There are no seas; liquid water cannot exist there, because the atmospheric pressure is too low. Unmanned space-craft which have made controlled landings there have shown no definite signs of life, but certainly Mars is much less unfriendly than any planet in the Solar System apart from Earth, and it is very likely that colonies will be established there during the first half of the coming century.

We need not say much about the asteroids, which are true dwarfs; only one (Ceres) is as much as 500 miles in diameter, and only one (Vesta) is ever visible with the naked eye. It seems certain that the planets were formed, around 4.6 thousand million years ago, from a cloud of material associated with the youthful Sun, but no planet of large size could form in the asteroid zone, because of the powerful disruptive pull of Jupiter.

Jupiter and Saturn are the giants of the Sun's family. They are very different from Earth; they are believed to have hot, rocky cores, surrounded by layers of liquid hydrogen which are topped by cloud-laden atmospheres. Uranus and Neptune are rather different in composition, but they too have gaseous surfaces, and certainly there can be no thought of attempting a manned landing on a giant planet. Most of our knowledge about these two outer worlds comes from one space-craft, *Voyager 2*, which by-passed Uranus in 1986 and Neptune in 1989 before starting on a never-ending journey out of the Solar System. Unfortunately no probe has yet been sent to Pluto, which is in a class of its own; it is bitterly cold, with a thin atmosphere, and it is accompanied by a secondary body Charon, which has more than half the diameter of Pluto itself.

The Moon

Apart from Mercury and Venus, all the planets are attended by satellites. We have one, our familiar Moon, which also depends on reflected sunlight. It is perhaps surprising to find that the Moon is by no means a good "mirror"; it sends back less than 10 per cent of the sunlight which falls upon it, so that it is by no means "shiny". Its sends us very little heat, so that there is no danger in looking straight at it through a telescope – though, as we will stress later in this book, direct telescopic observation of the Sun is emphatically not to be recommended.

The Moon shines so brilliantly in our skies that it is sometimes hard to credit that it is a very junior member of the Solar System. On average it is less than a quarter of a million miles away, and it is our faithful companion in space, staying together with us as we journey round the Sun. It is usually said that "the Moon moves round the Earth". To be more precise, the Earth and the Moon move together around their common centre of gravity, which is known as the barycentre – but since this point lies deep inside the Earth's globe, the refinement is not important to anybody except a mathematician.

The Moon takes 27.3 days to complete one orbit. Everyone must be familiar with its phases; Fig. 1.1 shows why they occur. When the Moon is at Position 1, its dark side faces us, and we cannot see it at all except when it passes directly in front of the Sun; this is the true "new moon". As it moves along it begins to show up as a crescent, and when it reaches Position 2 half of it can be seen; rather confusingly, this is known as First Quarter (because one-quarter of the orbit has been completed). Between Positions 2 and 3 the Moon is "gibbous", and when it reaches 3 it is full, with the whole of the sunlit hemisphere turned in our direction. Between 3 and 4 it is again gibbous; half at 4 (Last Quarter) and then a crescent once more before returning to new at Position 1.

There is one slight complication. The Earth and the Moon are travelling together round the Sun, and this means that the interval between one new moon and the next is not 27.3 days, but over 29 days (see Fig. 1.2); this period is known as a lunation, or synodic month. Note too that when the Moon is in the crescent phase, the "dark" side can usually be seen shining dimly. There is no mystery about this: it is due to light reflected on to the Moon from the Earth, and we call it the Earthshine. Country folk refer to it as "the Old Moon in the Young Moon's arms".

The Moon's path around us is not completely circular, as we will see later. It is also inclined at an angle of

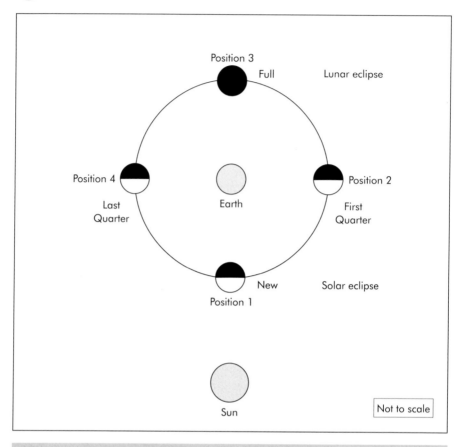

Figure 1.1. The phases of the Moon.

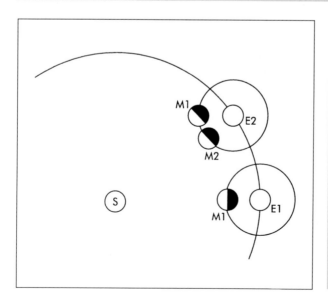

Figure 1.2. The lunation. In this figure, the Earth is at E1; the Moon is full, at M1. After 27.3 days the Moon has completed one orbit, and has reached M1 once more (see Fig. 1.1), but the Earth has moved on from E1 to E2. The Moon has to move on, to position M2, before it is again full; this takes over 2 extra days, so that the length of the lunation is 29d 12h 44m 3s.

just over 5 degrees to the orbit of the Earth. If the two paths were in the same plane, there would be a solar eclipse at every new moon, but the slight tilt means that on most occasions the new moon passes unseen either above or below the Sun in the sky.

The Moon spins slowly on its axis. The rotation period is 27.3 days, exactly the same as the time taken to complete one orbit – so that with slight modifications, due to the orbital eccentricity, the Moon keeps the same face turned towards us all the time; until 1959, when the Russians sent a space-craft on a "round trip", we had no direct knowledge of the averted hemisphere. Again there is no mystery about this behaviour; it is due to tidal effects over the ages. Most other planetary satellites have similarly captured or synchronous rotations.

Though the Earth and the Moon are of the same age, they are very different in nature. This is mainly because the Moon is not only much smaller than the Earth, but is also less dense and less massive; place the Earth in one pan of a gigantic pair of scales, and you would need 71 Moons to balance it. This means that the pull of gravity there is relatively weak. Go to the Moon, as the *Apollo* astronauts have done, and you will find that you have only one-sixth of your Earth weight. This has one important consequence; it means that the Moon to all intents and purposes has no atmosphere.

Throw an object upward from Earth, and it will rise to a certain height, stop, and fall back. The faster you throw it, the higher it will go. If you could hurl it up at 7 miles (11 km) per second, it would never come down at all; the Earth's gravity would not be able to draw it back, and the object would escape into space. This is why this particular speed is known as the Earth's escape velocity. The air which you and I are breathing is made up of millions upon millions of molecules, all moving around quickly. If they could attain escape velocity, they would be lost. Fortunately they cannot do so, but things are very different on the Moon, where the escape velocity is a mere 1.5 miles (2.4 km) per second. Any air that the Moon may once have had has long since leaked away, so that today the lunar world is airless, waterless – and lifeless. There are no mists, no clouds; everything is always sharp and clear-cut.

Even with the naked eye you can see great dark patches on the Moon. These were once thought to be seas, and were given romantic names such as the Mare Imbrium (Sea of Showers), Oceanus Procellarum (Ocean of Storms) and Sinus Iridum (Bay of Rainbows).

We now know that there has never been any water in them, though they were once oceans of lava which welled up from inside the Moon.

There are mountains, hills ridges and valleys, but the whole lunar scene is dominated by the craters. They range from huge enclosures capable of holding an entire city down to tiny pits so small that from Earth they cannot be seen at all; some have terraced walls and massive central mountains, while others are low-walled and flat-bottomed. A few craters are the centres of systems of bright streaks or rays, which extend for hundreds of miles and are spectacular when the Moon is near full. Any telescope will give superb views of the Moon – though full phase, when there are almost no shadows, is the very worst time for the newcomer to start observing. A crater is at its most spectacular when on or near the terminator (that is to say, the boundary between the sunlit and the night hemispheres), since at that time its floor will be wholly or partly filled with shadow.

In 1651 an Italian astronomer, Riccioli, drew a map of the Moon and named the craters after important people, usually scientists; his system has been extended and followed ever since. Ptolemy, Copernicus, Newton, Halley, Einstein – all have their craters, as well as some rather unexpected people – notably Julius Caesar, not because of his military prowess but because of his association with calendar reform. There is even a crater called Hell, though this does not indicate any exceptional depth; it was named after a Hungarian astronomer, Maximilian Hell.

It was in July 1969 that Neil Armstrong and Buzz Aldrin stepped out onto the bleak rocks of the lunar Sea of Tranquillity. Nobody has ever bettered Aldrin's description of the landscape: "Magnificent desolation." There is no wind, no weather, no sound; little has happened on the Moon for several thousands of millions of years. The sky is black even during the long day, with the Sun and the Earth shining down. Until now, the Moon has never known life, but it will be strange if we do not have a permanent base there before the new century is very far advanced.

Other Moons

The satellite system of other planets are very diverse. Mars has two tiny attendants, Phobos and Deimos,

Table 1.2. The satellites in the Solar System

Satellite	Primary	Longest Diameter	
		miles	km
Ganymede	Jupiter	3274	5268
Titan	Saturn	3201	5150
Callisto	Jupiter	2987	4806
Io	Jupiter	2275	3660
Moon	Earth	2160	3476
Europa	Jupiter	1945	3130
Triton	Neptune	1681	2705
Titania	Uranus	981	1578
Rhea	Saturn	950	1528
Oberon	Uranus	947	1523
Tapetus	Saturn	892	1436
Charon	Pluto	753	1270
Umbriel	Uranus	727	1169
Ariel	Uranus	720	1158
Dione	Saturn	696	1120
Tethys	Saturn	650	1046
Ceres	The largest asteroid	584	984

which are less than 20 miles across and are probably captured asteroids rather than bona-fide satellites; Jupiter has four very large satellites and a swarm of smaller ones; Saturn has one major satellite, Titan, which has a thick nitrogen-rich atmosphere. Altogether there are sixteen planetary satellites with diameters of over 600 miles (100 km) and these are detailed in Table 1.2.

Note that Ganymede and Titan are larger than the planet Mercury, and all the satellites in the list down to and including Triton are larger than Pluto. Many of these satellites could produce total solar eclipses as seen from their primary planets – a point to which we will return later.

Comets

Comets are the vagabonds of the Solar System. They are not solid, rocky bodies; they have been called "dirty ice-balls", which is by no means a bad description of them. Most of them move round the Sun in highly elliptical orbits, so that we can see them only when they are moving in the inner part of the Solar System.

With one exception, really brilliant comets take so long to go round the Sun that we see them only once

over periods of many lifetimes, and they cannot be predicted. They have been rare during our own century;
there was a spectacular visitor in 1910 (the Daylight
Comet) but then a relative dearth; though in 1996–97
there were two, Hyakutake and Hale–Bopp. (In general,
comets are named after their discoverers.) They are
believed to come from a cloud of icy bodies, known as
the Oort Cloud, orbiting the Sun at a distance of at least
a light-year. Other comets have much shorter periods,
so that they are seen regularly; it is now believed that
these come from a much closer cloud of bodies, not too
far beyond the orbits of Neptune and Pluto.

The only substantial part of a comet is its nucleus –
never more than a few miles across – which is made up
of ice together with "rubble". When in the far part of its
orbit, a comet is inert, but when it draws inwards and
is warmed by the Sun's rays the ices begin to evaporate,
and the comet may produce a head and a tail or tails.
The only bright comet which comes back regularly is
Halley's, named after Edmond Halley, the second
Astronomer Royal, who was the first to compute its
orbit. It has a period of 76 years; it was spectacular in
1835 and 1910, but was badly placed for its return of
1986, though it was easy to see with the naked eye. It
will be back once more in the year 2062. Unmanned
space-craft sent to it in 1986 showed a darkish nucleus
shaped rather like a peanut.

Comets used to cause alarm, but in fact they are
flimsy bodies. It is true that a direct hit on Earth would
cause widespread devastation, but our world is a small
target, and the danger of a major collision is very
slight.

Meteors

As a comet moves along it leaves a dusty trail behind it.
When the Earth passes through such a trail, particles
dash into the upper air, become heated by friction, and
burn away to produce the luminous streaks known as
meteors or shooting-stars; they end their journey to the
ground in the form of very fine dust. There are many
regular displays, most of them associated with known
comets; the best shower is that in August, which seems
to come from the direction of the constellation Perseus.
There are also "sporadic" meteors, which may appear
from any direction at any moment.

Meteorites are larger bodies, which may land intact and produce craters – such as the famous Arizona Crater, a popular 50,000-year-old tourist attraction. Meteorites are not associated with comets or with shooting-star meteors; they come from the asteroid belt, and there is no distinction between a large meteorite and a small asteroid. Again there is always the chance of a direct hit, and it has been suggested that the dinosaurs were wiped out, around 65,000 years ago, by a climatic change produced by the impact of a meteorite, asteroid or comet. This is certainly possible, though it is no more than a plausible theory.

The Solar System is a fascinating place – ruled by the Sun, its other members are very minor in comparison. Without the Sun, the planets would never have come into being – and neither would we.

Chapter 2

Daytime Star

The Sun is immense. Its diameter is over 100 times that of the Earth, and it is 333,000 times as massive. It is also very hot. Even at the surface the temperature is between 5000 and 6000°C, and near the core, where the Sun is producing its energy, the temperature rises to a staggering 15,000,000°C.

There is no solid surface to the Sun; it is gaseous all the way through its globe, and its outer layers are tenuous, though at its centre the density is at least thirteen times as great as that of lead. Of course the Sun is not burning in the manner of a coal fire. If it were made up of ordinary coal, radiating as fiercely as the Sun actually does, it would not last for very long before being turned into ashes – yet we know that the Earth is over 4000 million years old, and the Sun is certainly older than that. Astronomers rate it as a dwarf star; it is in a stable state, and will not change much for some thousands of millions of years in the future, though its life-span is limited, and eventually it is bound to destroy the Earth.

The bright surface which we see is termed the photosphere, and it is upon this photosphere that we see the famous dark patches which we call sunspots. (The only sensible way to study the Sun telescopically is by the method of projection; we will say much more about this later.) Sunspots are not always on view, and there are times when the photosphere appears completely blank, but at other times the surface is very active indeed.

Sunspots

A large sunspot is made up of a dark central portion or umbra, surrounded by a lighter penumbra; sometimes the shapes are regular, sometimes almost incredibly complex, with many umbrae contained in the same penumbral mass. Individual sunspots are common enough, but more generally they are appear in well-marked groups. Each group has its own characteristics, but on average has two spots seen as a pair of tiny pores at the limit of visibility; these develop into proper spots, growing and separating in longitude. Within a fortnight the group reaches its maximum length, with a fairly regular leading spot and a less regular follower – together with many minor spots and clusters. Then a decline sets in, and eventually the group disappears; in most cases the leader is the last survivor.

Large groups cover a wide area; the record is held by the group of April 1947, which at one time spread over 18,000 million square kilometres (7000 million square miles) and lasted for several months. Small spots may vanish altogether after only a few hours.

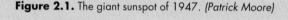

Figure 2.1. The giant sunspot of 1947. *(Patrick Moore)*

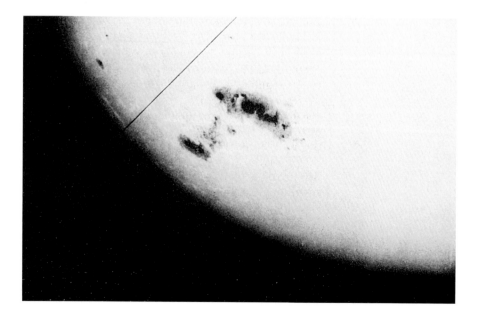

The term "spot" may be misleading. As long ago as 1774 the Scottish astronomer Alexander Wilson noted that with a regular spot, the penumbra to the limbward side of the Sun is broader than the penumbra on the opposite side, indicating that the spot is a hollow rather than an elevation. Not all spots show the Wilson effect, but in many cases it can be very obvious.

Sunspots are not really dark; they appear so only because their temperatures are around 2000 degrees cooler than that of the surrounding photosphere. If a spot could be seen shining on its own, the surface brilliancy would be greater than that of an arc lamp. This is well shown on the rare occasions when Mercury and Venus pass in transit. The disks of the planets appear jet-black, and the contrast with any spots which happen to be on view is striking.

If a spot or spot-group is observed from day-to-day, it is seen to be carried across the disk by virtue of the Sun's rotation. From this, it is easy to work out just how the Sun spins on its axis. Near the equator, the rotation period is about 25 days, but the value in higher latitudes is much longer, and near the solar poles it is over 30 days. In fact, the Sun shows what is termed differential rotation, and does not behave in the way that a solid body would.

In the mid-nineteenth century a German amateur, Heinrich Schwabe, made daily observations of the Sun, and realised that there is a well-defined cycle of activity. Every 11 years or so the Sun is at its most energetic, with many spots and groups; activity then dies down, and at solar minimum there may be many days with no spots at all, after which activity builds up again toward the next maximum. The cycle is not perfectly regular, and the 11-year period is only an average; neither are all maxima equally energetic. The last maxima occurred in 1980 and 1991, so that the next may be expected early in the new millennium; the last minimum took place in 1996.

It is interesting to note that according to all the evidence, spots were almost absent from 1645 to 1715, suggesting that the cycle was suspended – for reasons which we do not know. This was pointed out by the English astronomer E.W. Maunder from a study of old records, and the period has now become known as the Maunder Minimum. Whether another spotless period will occur in the future remains to be seen. Activity is often given by what is termed the Wolf or Zürich number, given by the formula:

$$R = k\,(10g + f),$$

where R is the Zürich number, g is the number of groups seen, and f is the total number of individual spots, while k is a constant depending on the equipment and site of the observer (in practice, k is usually not very different from unity). The Zürich number may range from 0 for a clear disk to well over 200. The graph given here (see Fig. 2.2) is based on observations by one of us (PM) between 1984 and 1997; this maximum was not particularly energetic.

Another German, F.G.W. Spörer, drew attention to a law relating to the latitudes of the spots. At the start of a new cycle, following minimum, the first spots appear at between latitudes 30 and 45 degrees north or south. As the cycle progresses, groups break out closer to the equator, until at maximum the average latitudes of the spots are no more than 15 degrees north or south. After maximum the spots become less common, but the approach to the equator continues down to about 7 degrees north or south. The spots of the old cycle then die out, but even before they have completely disappeared the first spots of the new cycle appear at higher latitudes. Note that individual spots and groups do not drift around in latitude – and that spots never occur at the solar equator or the poles.

It cannot be said that we have a complete understanding of sunspots, but they are certainly associated with magnetic phenomena. The accepted theory was

Figure 2.2. Graph of solar activity (Zürich number) against time.

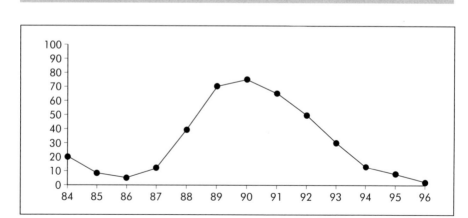

proposed by H. Badcock in 1961. It is assumed that the Sun's lines of magnetic force run underneath the surface from one magnetic pole to the other. When they break through to the photosphere, they cool and calm it, so that spots appear. In a two-spot group the leader and the follower will have opposite magnetic polarities – that is to say, if the leader is a "north polarity" spot the follower will be "south". Conditions in the opposite hemisphere are reversed, and after two cycles there is a complete reversal, so that there are grounds for supposing that the true length of a cycle is not 11 years, but 22.

Even in non-spot zones, the Sun's surface is not quiescent. The photosphere has a granular structure; each granule is about 600 miles (1000 km) in diameter, and has a lifetime of about 8 minutes, so that at any one time the whole surface involves about 4,000,000 granules. The general situation has been compared with the boiling of a liquid, though of course the photosphere is purely gaseous. Super-granules involve large organised cells, usually of the order of 19,000 miles (30,000 km) across, which contain hundreds of individual granules. Material wells up in the centre of the cell, spreading out to the sides before sinking again.

We have also found that the visible surface is oscillating, though not by more than about 80 ft (25 m). Apparently this is due to pressure waves ("sound") which echo and resonate in the Sun's interior. A wave originating deep inside the Sun is bent or refracted to the surface, because the speed of the wave is greater at greater depths. On striking the surface and rebounding back downward, the wave makes the chromosphere oscillate. There are many "oscillation modes", and the whole situation is much more complicated than we believed only a few years ago.

Faculae (Latin: torches) are bright areas, consisting of incandescent hydrogen, lying just above the photosphere and are usually associated with spots; they may also be seen in places where a spot is about to break out, and they may persist for some time after a spot has disappeared. Flares are violent, short-lived phenomena, associated with active spot groups; they emit charged particles as well as short-wave radiation, and do not last for more than a few hours at most. They are seldom seen in ordinary light, and are therefore studied spectroscopically.

Spectroscopy

Just as a telescope collects light, so a spectroscope splits it up. Remember, light is a wave-motion; the colour depends on the wavelength, from red at the long-wave end of the range through orange, yellow, green and blue down to violet at the short-wave end. The splitting-up is achieved by using a glass prism or some equivalent apparatus; the red component will be bent or refracted the least, violet the most. An incandescent solid, liquid, or gas at high pressure will produce a rainbow or continuous spectrum, while an incandescent gas at lower pressure will yield isolated bright lines. Each of these lines is due to some particular element or group of elements, and cannot be duplicated by any other substance. For example, the spectrum of sodium (one of the constituents of common salt) includes two bright yellow lines; once these are seen, it follows that sodium must be responsible.

In 1814 the German optician Josef Fraunhöfer made careful studies of the Sun's spectrum. He passed sunlight through a slit, and then through a prism, and observed a rainbow band crossed by dark lines – still often called the Fraunhöfer lines (though he was not the first to see them; they had been observed in 1802 by W.H. Wollaston, in England). Fraunhöfer found that the lines were permanent, always in the same positions and of the same intensities. Years later, a correct explanation was given by two of Fraunhöfer's countrymen, G. Kirchhoff and G. Bunsen. The solar photosphere produces a continuous spectrum, superimposed on which are lines produced by the more rarefied gases lying at higher levels; because the lines are seen against a brighter background, they appear dark instead of bright, but otherwise they are unaltered, so that they can be identified. Most of the known chemical elements have now been identified in the Sun by their unique spectral "fingerprints", but of course the spectrum is very complex; iron alone accounts for many thousands of absorption lines. This leads us on to a knowledge of the Sun's make-up.

The Sun's Composition

There were, of course, many early theories which attempted to explain the energy-generating mechanisms in the Sun, but all suffered fatal objections.

One of the first was due to Hermann von Helmholtz and independently by William Thomson (later Lord Kelvin) in the middle of the nineteenth century. To them, the Sun was heated up by gravitational energy. The idea is simple enough; if anything falls through a gravitational field, the loss of kinetic energy appears as heat. It only needs the Sun to contract by around 100 feet per year to create enough heat to last 50,000,000 years. This was ample for the geological theories of that age, but there was a problem which soon became painfully evident: the Earth was much older than 50,000,000 years. Darwin's theories and the geological evidence showed that. There had to be some other process. Radioactivity appeared promising, but it was then found that there was not enough uranium and other radioactive elements in the Sun to maintain it for a sufficient length of time.

The problem was finally solved in the 1930s, independently by Hans Bethe in America and Carl von Weizsäcker in Germany. It was they who realised that the essential "fuel" is hydrogen.

A few comments about atomic theory are needed here. The atom of hydrogen is the lightest of all, and it makes up a great deal of the mass of the Sun – indeed, over 70%. If four hydrogen nuclei are combined to form one nucleus of the second lightest element, helium, a little mass is lost and a little energy is set free. It is this energy which keeps the Sun shining, and the mass-loss amounts to 4,000,000 tons per second (see display panel overleaf). This may seem a great deal, but it is not much when compared with the total mass of the Sun. It is now clear that the Sun has been shining for at least 4000 million years, and is only middle-aged.

There are various methods in which this conversion may be achieved. The details are rather beyond our scope here – the interested reader should refer to the boxed text on pages 20 and 21 – but the end result is the same: hydrogen is changed into helium, and the Sun continues to shine. It is, in fact, a vast nuclear reactor, and it emits across the whole range of wavelengths or "electromagnetic spectrum". Eventually it will use up its available hydrogen fuel, and will change its structure; the outer layers will swell out and cool, so that the Sun will become a red giant star – and will certainly destroy the Earth. It will then throw off its outer layers, and end up as a very small, dense star of the type known as a white dwarf. We, however, will not be there to see it.

There is still a great deal about the Sun that we do not know. For example, theory says that it sends out vast numbers of particles known as neutrinos, which are far from easy to detect because they have little mass (perhaps none at all) and no electric charge. Experiments show that there are far fewer solar neutrinos than there ought to be, and we do not know why. The generally-accepted temperature of the Sun's core is 15,000,000°C; if we reduce this significantly we can account for the relative dearth of neutrinos, but this introduces all manner of other theoretical problems.

The Sun sends us great numbers of cosmic rays (which are not rays at all, but atomic particles) which disrupt the Earth's magnetic field. It also seems that the Sun has a layered structure, with two different energy reactions occurring at two different depths. Granulations, sunspots, and even prominences can be likened to the phenomena in a boiling liquid – though, as we have seen, the Sun's surface is purely gaseous.

Nuclear Reactions in the Sun

The main energy source of the Sun is provided by the proton–proton reaction (one proton is the same as a hydrogen nucleus). Two protons react to form a deuteron and other particles (a positron, the positive opposite of an electron, and a neutrino). The positron is at once annihilated by combining with an electron, emitting a burst of very short-wave gamma radiation. The deuteron then reacts with another proton to produce a form of helium known as helium-3, and more gamma radiation. This process has to occur twice before the next step can take place; two helium-3 nuclei collide to form a single helium-4 nucleus and two recreated protons, with more gamma radiation. The process may be detailed as follows (H standing for hydrogen, He for helium, e for electron, ν for neutrino):

$$^{1}H \; + \; ^{1}H \; \rightarrow \; ^{2}H + \qquad e^{+} + \nu$$
$$e \; + \; e^{+} \; \rightarrow \; \text{gamma energy}$$
$$^{2}H \; + \; ^{1}H \; \rightarrow \; ^{3}He + \qquad \text{gamma energy}$$
$$^{3}He \; + \; ^{3}He \; \rightarrow \; ^{4}He + \; ^{1}H + ^{1}H + \text{gamma energy}$$

In stars like the Sun, the reactions can only occur to any appreciable extent at the very centre where the pressures and temperatures are high enough. We need to be in the 8 million K range for this to start. We now think that at least two other variants of the proton– proton reaction can occur, one involves beryllium and lithium (about 8%) and another boron and beryllium (a mere 1%).

The really important alternative carbon–nitrogen cycle needs some carbon and nitrogen to work, and as these are formed after a supernova, the energy source is not important in the very first generation of stars. The cycle also needs higher temperatures, around 20 million K, making it more likely in the deeper regions of stars larger than the Sun. The energy released is slightly less than the proton–proton cycle at 4.0×10^{-5} erg, this is because the second neutrino carries the missing energy into deep space.

Carbon is converted to nitrogen, then to oxygen, before being recreated. The process is more efficient, being the predominant energy source in massive stars.

$$^{12}C + \quad ^1H \quad \rightarrow \quad ^{13}N + \text{gamma energy}$$
$$^{13}N \quad \rightarrow \quad ^{13}C + e^+ + v$$
$$^{13}C + \quad ^1H \quad \rightarrow \quad ^{14}N + \text{gamma energy}$$
$$^{14}N + \quad ^1H \quad \rightarrow \quad ^{15}O + \text{gamma energy}$$
$$^{15}O \quad \rightarrow \quad ^{15}N + e^+ + v$$
$$^{15}N + \quad ^1H \quad \rightarrow \quad ^{12}C + ^4He$$

The striking difference from anything with which we are familiar on Earth is that the "boiling" process is not rapid. It seems that at least a million years must elapse before energy created near the core reaches the surface, and has been transformed into visible light. If we could imagine that the Sun stopped its energy production a few hundred thousand years ago this energy would still have a long way to go to reach the surface.

The huge regions of hot gases ejected at the Sun's surface can be regarded as pointers to the internal conditions. By studying these surface phenomena at times of total eclipses, we may hope to form a better understanding of the processes going on deep inside the globe.

The Structure of the Sun

From Earth, the Sun's surroundings can be seen with the naked eye only during a total eclipse. For many years it was not known whether they were due to the Sun itself, or to an extended lunar atmosphere; only a hundred years ago it was finally proved that they belong to the Sun.

Immediately above the photosphere lies the chromosphere, or "colour-sphere". At the eclipse of 1851 it was described in detail by Sir George Airy, who called it the "sierra", because he believed that its jagged upper edge might be due to solar mountains; only when it became clear that the surface is wholly gaseous was the term "chromosphere" introduced. The main constituent is hydrogen, which explains the redness. It is

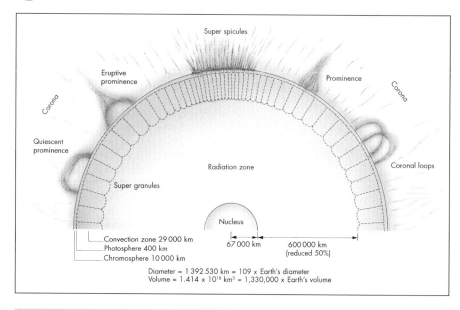

Figure 2.3. Standard model of the Sun.

composed of rarefied gases, and it is here, and in the uppermost part of the photosphere, that the dark absorption lines in the solar spectrum are produced – hence the term "reversing layer". The chromosphere is from 1250 miles (2000 km) to 6000 miles (10,000 km) deep, and the temperature rises from 4200°C at the base up to 8000°C at an altitude of 900 miles (1500 km), where the chromosphere merges with the lower corona.

Note here that the scientific definition of "temperature" is not the same as the everyday meaning of "heat". Temperature depends upon the speeds at which the atoms and molecules are moving around; the greater the speeds, the higher the temperatures. Thus the upper chromosphere is at a higher temperature than the base, because the speeds of the particles are faster, but there is little heat, because there are so few particles. There is an analogy here with a glowing poker and a firework sparkler. Each spark of firework is at a high temperature, but is of such low mass that hand-holding is quite safe; a glowing poker is at a much lower temperature, but neither of the present writers would care to take hold of it!

Just before totality, the chromosphere is shining "on its own", and the dark absorption lines change abruptly into emission lines; this is termed the flash spectrum. C.A. Young gave the first vivid description of it, at the eclipse of 1870, which he observed from Spain:

> As the Moon advances, making narrower and narrower the remaining sickle of the solar disk, the dark lines of the spectrum remain for the most part sensibly unchanged, though becoming somewhat more intense. A few, however begin to fade out, and some even begin to turn palely bright a minute or two before totality begins. But the moment the Sun is hidden, through the whole length of the spectrum as in the red, the green, the violet – the bright lines flash out by hundreds and thousands almost startlingly; as suddenly as stars from a bursting rocket head, and as evanescent, for the whole thing is over in two or three seconds. The layer seems to be only something under as thousand miles of thickness, and the Moon's motion covers it very quickly.

The Prominences and Corona

Prominences are phenomena of the solar atmosphere; they were once termed "red flames", and during totality they do indeed look flamelike. They are made up mainly of hydrogen, and at the eclipse of 1868 the French astronomer Jules Janssen realised that it would be possible to observe them at any time, without waiting for an eclipse. He set the slit of his spectroscope at the limb of an image of the Sun, and saw the bright hydrogen emission line in the same position where a prominences had been during the eclipse. From this he was able to keep the prominences under surveillance, and to draw them as they changed in form. The English astronomer Norman Lockyer came to the same conclusion quite independently, and he too was able to follow the prominences continuously – as is done today by all serious solar observers, both amateur and professional. By observing at hydrogen wavelengths, prominences may be seen against the photosphere as dark filaments, termed flocculi; bright flocculi are due to the presence of calcium.

Active or eruptive prominences change in form very quickly; there are vast "loops", and very often material is hurled away from the Sun altogether. Quiescent prominences may hang in the solar atmosphere for months. Obviously there is greatest activity near spot-maximum, and there are also greater numbers of the violent, short-lived flares.

The upper chromosphere merges into the Sun's outer atmosphere, the corona, which is made up chiefly of hydrogen. (An unknown element was once reported spectroscopically, and was even given a name – coronium – but it was later found out that the lines were due to ionised iron, nickel and calcium.) The density of the corona is less than one million millionth of that of the Earth's air at sea-level, and it sends us only about one-millionth as much light as the photosphere, which is why direct observation during totality is safe. At times of non-eclipse it is very hard to study from Earth. Instruments known as coronagraphs have been able to show the innermost part, but to see the corona properly one must either wait for a total eclipse or else for a trip into space, above the Earth's atmosphere.

The corona has no definite boundary, but simply thins out with increasing distance from the Sun until its density is no greater than that of the interplanetary medium. The temperature is very high indeed, and amounts to several millions of degrees. All sorts of explanations have been offered – sound-waves from below, magnetic effects and so on – but we have to admit that we are still very unsure of the reason for this curious state of affairs. Of course, the amount of "heat" sent out from the corona is negligible.

It has long been known that the Sun sends out a continuous flow of charged particles, termed the solar wind; it consists of plasma, a mixture of protons, helium nuclei and electrons. It was first identified in 1951 by L. Biermann, who realised that it was causing the gas tails of comets to point away from the Sun; the solar wind particles repel the tiny pieces of dust and drive them outward from the comet's nucleus, so that when the comet is travelling away from the Sun it moves tail-first. (The dust tails of comets also point away from the Sun, but this is because of the slight but appreciable pressure of sunlight.) The velocity of the solar wind ranges from 125 miles (200 km) per second up to 550 miles (900 km) per second; low-velocity streams come from loops in the corona, but higher-velocity streams come from "coronal holes", regions of lower density in the corona. These holes are most evident near the solar poles, where the magnetic field lines are open and the charged particles can escape with relative ease. We can never have a proper view of the Sun's poles from Earth, because our view is always more or less broadside-on, and so in 1990 a special probe, *Ulysses*, was launched to observe them.

Ulysses moved well out of the ecliptic plane (using the powerful pull of Jupiter to put it into the desired orbit) and has been a great success; it has shown that the magnetic properties of the Sun are very different from anything which had been expected, and the magnetic poles are much less well-defined.

The shape of the corona varies according to the state of the spot-cycle. Near minimum the corona is comparatively symmetrical, while long streamers are seen near spot-maximum. Incidentally, it seems that during the Maunder Minimum of 1645–1715 the corona was very inconspicuous, though unfortunately the records of the eclipses at that period are very fragmentary.

The solar wind pervades the entire Solar System. At a sufficiently great distance it will cease to be detectable; this boundary is known as the heliopause, enclosing the region, termed the heliosphere, in which the Sun's influence is dominant. It is hoped that some of the deep-space probes (*Pioneers 10* and *11*, and *Voyagers 1* and *2*, dispatched to the outer planets) will remain in contact long enough to reach the heliopause, and send back some definite information about conditions there.

From space-craft, of course, the corona can always be seen, and at present there are various solar probes in orbit around the Sun – notably SOHO (the Solar Heliospheric Observatory) launched in December 1995. Therefore, it cannot be denied that the importance of eclipses has been somewhat lessened. Yet there are still many branches of research which can be carried out only when the Moon hides the Sun; and, in any case, who would pass over the chance of seeing the splendour of a total eclipse?

The more serious observer will want to make accurate plots of the spot positions, and to measure the areas of the groups; for this special disks are used. But for the casual observer, the main concentration is upon sketching – and of course photography.

Chapter 3

Observing the Sun

Look at the Sun when it is low over the horizon, and shining through a layer of haze or mist. It seems beautiful – and it is. It also seems safe – which it most emphatically is not. Even when it appears deceptively harmless, the Sun is dangerous. Staring straight at it is not to be recommended, and to look at it direct through an unfiltered telescope or pair of binoculars is certain to result in permanent eye damage, probably total blindness. Serious damage can be done in a fraction of a second.

This is not mere alarmism; there have been tragic accidents in the past, particularly just before or during an eclipse, when members of the general public are aware that something unusual is happening, but are not alive to the Sun's dangerous nature. Do you recall the old Boy Scout method of starting a camp-fire by using a piece of glass to focus the Sun's rays on to straw or dry grass? It does not need much in the way of imagination to picture what would happen if this intensely hot spot were concentrated on the retina.

Around the time of solar maximum, there are occasional spots and spot-groups which are large enough to be seen with the naked eye, mainly when the Sun is veiled by mist. This is a particularly hazardous period, because of the temptation to take a telescope or pair of binoculars and "have a closer look". There are ways of reducing the danger, but the greatest care is always needed.

Pinhole Projection

It is possible to use the simplest of all optical instruments to project a solar image – the pinhole.

The most elementary of these can be made by pushing a needle through a tinplate, or by using a drill. The hole can be of almost any diameter, and a few quick trials will sort out the best one for the audience size. During eclipses many natural pinholes are "pressed" into service and one of the most striking sights during the partial phases is the multitude of crescents cast on the ground under trees (Fig. 3.1).

The principles of the optics are the same as with lenses, and the Sun's image size is approximately 1/100 that of the distance from the hole to the screen. The idea of using pinholes has been around for a long time and the "camera lucida" was a favourite with many artists. The instrument might even have been the inspiration for the first cameras. These instruments worked best in a darkened space, and the other artist often covered himself in a dark cape very much like the early photographers studying their focusing screen. It was only a small step to view the Sun as its projected image on a wall in a darkened room, from a chink in the blind.

Before the invention of telescopes, pinhole projection was the only way of viewing an enlarged image of the Sun. The first true astronomers of the middle ages, such as Kepler, used the scheme to observe eclipses and sunspots.

An interesting variant of the pinhole in a darkened room is the camera obscura. Are there any already on the track of the 1999 eclipse?

Figure 3.1. The pinhole effect of tree leaves. Arizona, 1994. (Michael Maunder)

Filters for Solar Observation

Extreme caution is needed here, because we are born with only one pair of eyes. The Sun is a phenomenally powerful nuclear reactor, and a major part of the nuclear fallout is the infrared radiation which keeps us warm on this planet. Do not be stupid enough to try to view the Sun directly through any form of optical aid. The heat will literally fry your retina, and medical science has not advanced to the point where sight can be restored in a fried eyeball. Damaged eyesight is the best you can hope for or, more likely, blindness will be the result.

There are some very simple rules to obey when observing the Sun when not in total eclipse.

Safety Warning on Filters

1. Be on your guard all the time.
2. Whenever possible, use some alternative method such as projection. This warning cannot be repeated often enough.
3. No filter placed in line with the main mirror or lens can be relied on to be 100% safe.

All these rules are important, but what happens if you have to use a filter? Rule 3 overrides. Never, ever, consider a filter behind the main optics, no matter whatever the source, or type. Things can, and do, go wrong in real life.

An Alternative

There is an alternative to solar projection which could be regarded as a "fudge" for a proper filter. This is reflection.

A very small proportion of light is reflected from unsilvered glass, and this has been used quite successfully in some forms of prominence telescopes. Reflections from two surfaces are usually sufficient to reduce the heat energy to acceptable levels, as the bulk passes straight through the glass into the wide yonder. If the rest of the optics are of a high focal ratio, in specially designed equipment, the heat hazard can be reduced to manageable proportions. Whilst quite an effective method and widely used at one time in professional equipment, it is quite difficult to keep track of the image orientation. The Sun's image is a disk, and it is never easy to determine the orientation after reflections in normal circumstances. Some guidance can be gained by allowing the image to "trail" on an undriven polar mount.

A new problem has arisen in recent years in still and video cameras relying on the right sort of polarisation for autofocus. Where this is found to be a problem, there is no alternative to using a conventional filter, or a manual camera/video.

Some of the uncertainties in image orientation can be removed by using a single "black" mirror. Black glass also reflects a small portion of the light, and if well chosen will also absorb most of the heat. Welder's glass is one option, and is available in large optically flat slabs.

The great advantage of this single mirror is that several cameras can be focused on it at once with minimal distortion just "off-axis". A neutral density filter of some sort is still needed on the lens for complete safety, but it does not have to be as dense as the more specialised ones described next. The main requirement here is that the safety filter cuts out residual heat. The idea worked well at the 1973 eclipse observed from a ship, the *Monte Umbe*. A special mount was made to house the various still and ciné cameras, and just at totality, the black mirror was flipped over to be replaced by a normal mirror, so that everything remained in focus (see Fig. 3.2).

The major disadvantage with this idea was that the image "flipped" from left to right, and whilst this can be allowed for in final processing, great skill and practice was needed to track the solar image in reversed-motion on a moving boat. Working on land would be easier – just make copious notes to correct for the lateral inversions introduced. Some still cameras run the film in a contrary direction.

Filter Types

Filters mounted in front of the main optics must have the right properties to remove invisible ultraviolet

Figure 3.2. *Monte Umbe* (Mauretania), 1973. Norman Fisher, Keith Brackenborough and Margaret Rumsby demonstrate black glass reflection technique. *(Michael Maunder)*

(UV) and infrared (IR) radiation. Use only solar filters that remove both UV and IR radiation.

UV radiation is not always considered in solar filter choice. It should be. UV exposure can lead to similar effects as sunburn; you are unlikely to absorb enough radiation to end up with cancer, but corneal cataracts are a strong possibility with prolonged usage. UV radiation passes through a surprisingly large number of filters which look completely opaque.

IR is even more prone to pass through visually opaque filters. IR is pure heat radiation. If concentrated on the retina or in the eyeball fluid, it will quite literally cook it. Irreversible vision loss occurs in the areas affected, and shows up immediately.

Mylar

Mylar filters have become the most popular solar filters in recent years, mainly because of their cheapness.

Mylar is an extremely tough proprietary plastic film a few microns thick, of amazing strength. Solar filters are made by coating the film with a very thin layer of aluminium metal. For visual work, the Mylar coated on both sides is by far the safest, whereas single-sided is better for photography because of the shorter exposure times this allows.

The aluminium coating absorbs both UV and the bulk of the IR radiation, and transmits only a small amount of visible light, mainly blue. Properly prepared and selected, Mylar filters are safe enough for everyday use. Many viewing "spectacles" are made with Mylar.

Mylar film must never be confused with "silver paper", a thin sheet of aluminium metal, nor with glitters. Glitters look like silver paper but are, in fact, made by printing silver or metallic ink on to a wide variety of plastics. Glitters are very dangerous indeed, because they are often made to imitate the real thing. They pass too much damaging radiation; use only Mylar from a reputable source.

The Mylar material itself is extremely tough, but the aluminium coating on its surface is not. Inspect for pinholes each and every time before you put your eye to the optical endpiece. Pinholes act like pinhole cameras and transmit a lot of damaging radiation. They also degrade image contrast and sharpness. Also inspect the surface regularly for unevenness. Scuffing and other abrasions eventually lead to local highspots which behave like pinholes before forming pinholes proper. The health dangers are obvious and image degradation can be appreciable, a useful warning sign itself. Some slight rucking or unevenness rarely leads to image degradation. On the contrary, some authorities positively recommend not stretching the film too tightly. If a pinhole hotspot is seen, replace the filter immediately – the stuff is cheap enough.

Mylar film has a marked polarisation (see Fig. 3.3), as you can see by holding a piece up to sunlight. The Sun's image will often have "wings", which rotate as the piece rotates.

The effect doesn't matter in visual work but has become serious with autofocus cameras operating with circular polarisation. Before attempting to photograph

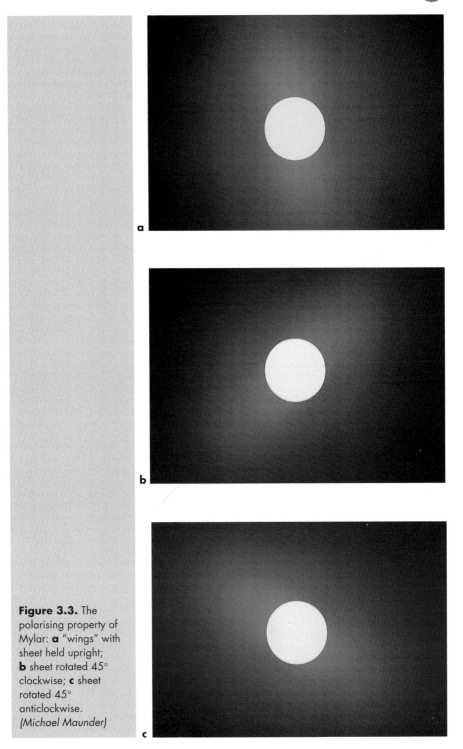

Figure 3.3. The polarising property of Mylar: **a** "wings" with sheet held upright; **b** sheet rotated 45° clockwise; **c** sheet rotated 45° anticlockwise. (Michael Maunder)

anything important, check focus by rotating the filter, then double check with test exposures.

Mylar film can be obtained in a very wide range of densities, and this leads to a few problems of its own. There is a tendency to use the lighter versions for photography and forget the visual dangers during focusing and regular visual monitoring. For safety, always use the denser products. However, there is considerable merit in keeping exposure times short, particularly with the longer focal lengths needed in partial and annular eclipses and for granularity and other studies in routine solar surveys.

Try the effect of two thicknesses for visual work and for focusing. Provided that the image is not too degraded, a point of best focus is quickly found. An alternative is to use a stop, though this might make the focus shift in "cheaper" optics.

Aluminised Glass

Regard these as a rigid version of Mylar. They are made in the same way, by vacuum depositing aluminium. Apply the same safety checks as for Mylar.

The "poling" problem should not happen. The important difference from Mylar is that the optical properties can be much more rigidly specified in manufacture and sale, but you pay accordingly. Many companies offer these filters for sale. The alternative is to get your own made by a company offering an aluminising service for normal astronomical mirrors. Choose one with the necessary quality control to ensure the absence of pinholes and hotspots, which are less critical in a conventional mirror. The company must have a track record of reliability.

Aluminised coatings can be laid down on virtually any size of glass capable of going into the vacuum chamber. The cost is invariably determined as a "unit" per run, so the limit is how many glass pieces will go into it. The greater the number of units, the cheaper each unit becomes. The glass must be of the highest optical quality, and sizes above 100 mm diameter are expensive. Fortunately, pieces bigger than this are not needed for most occasions. A popular camera "haze" filter size is 72 mm diameter, which covers many of the cheaper-end-of-the-market mirror lenses, ideal for cabin baggage.

Treat these filters with respect and they should last a lifetime. Many coating companies offer a sur-coating service, which extends the life considerably. Filters become suspect when they appear tarnished or stained as the aluminium oxidises. Avoid touching the surface with fingers, and also avoid any other contact with salt such as sea water, or corrosive atmospheres. Mount filters with the aluminised surface innermost, so that the glass itself acts as its own protection.

Like Mylar, the solar image tends to be a strong blue although some makers claim a neutral tinge. The colour is not important in monochrome work, but it is a distinct advantage with video with a facility to colour correct.

Inconel

This name has stuck for a variety of special stainless steels on optical glass. At present, these special filters are only readily available in the USA; most UK and European companies do not produce them. The technology has been around for a long time under a different guise as beam splitters and neutral density filters from many manufacturers.

The drawback is the cost. The reason is that these beam splitters and neutral density filters are the best normally available and are intended for very precise optical work. They are produced to extremely precise optical specifications and guarantees backed by International Standards' tests. Another problem is that virtually nobody makes these above 50 mm square, 25 mm being the industry standard. If this is big enough, use them because of their guaranteed pedigree.

Stainless steel filters are produced by at least one American company giving a six-year guarantee. The technology is now sufficiently mature for a "second generation" filter. This is claimed to be a triple layer of stainless steel on chromium on top of nickel. These new filters are guaranteed for 15 years, and are made as large as you can afford in a variety of densities for both visual and photographic work. Every solar observatory should aspire to own at least one of these, if only for the peace of mind given by a specified optical safety and durability.

It might be possible to find a company willing to coat to your own specification. Always choose chromium or a high percentage of it in the stainless mixture to ensure the absence of IR. Although much more durable than aluminium, pinholing is the usual fate for most of them. Do not rely entirely on the guarantee and always check.

Silver

All astronomical mirrors were silvered before aluminising technology took over. The latest generation of telescopes tends to have that metal coating because of its superior reflective power. Modern surface coatings ensure a respectable lifetime, and as the technology spreads some silver solar filters could appear in the amateur market. Standard textbooks on silvering describe all the old wet techniques, and it might be worthwhile dabbling.

Pure silver mirrors and neutral density filters are available from Balzers for beam splitting and as neutral density filters. These are specially coated for durability. They have the same drawbacks as Inconel – small sizes and high cost. The main advantage of a pure silver mirror as a filter is the perfectly neutral image colour – even better than Mylar in filtering out infrared radiation, and thereby safer.

"Black" Black and White Film

This is very much the "poor-man's" answer to silvering glass. Fully developed, fogged film is safe to use when dense enough, and it does give much better colours than Mylar. It is rarely used because the solar image generated through ordinary camera film is awkward to focus, and has low contrast due to halation arising from the Callier effect (light scattering by the silver particles in the emulsion).

It is possible to reduce the Callier effect with lith or line films, which have very thin and fine-grained emulsions of highest contrast. These films are also available in sheet sizes bigger than any normal solar telescope, and are made with a backing of Mylar. These films must be developed and fixed properly. Check them for pinholes, a characteristic feature of the film types, and reject any defective ones.

Another factor to keep in mind is that line and lith films should not be exposed to too much light before development, just enough to ensure complete fogging. Too much light and they can "solarise" which means that the final image density is bleached out and useless as a solar filter. Properly prepared, line and lith films give image densities as usable as Mylar film and Inconel. The great advantage is that the films make an excellent back-up if all else fails. A whole range of filters can be prepared with a range of densities to cover every conceivable cloud cover.

Monochrome film filters are definitely not "second best", and every solar eclipse traveller should acquire some. Their main use is as simple naked-eye visual filters. Mount them in ordinary slide mounts, and protect in thin polythene bags.

Welding Glass

This is an extremely useful material that is available in a very wide range of densities, called "shades". The direct viewing shades 13 and 14 are also right for most photography, although 12 is often preferred.

The stock shape is rectangular. Do not have them cut into circular or other shapes as this often generates stresses and strains which show up in the solar image. Check each sheet for optical flatness. For direct viewing, flatness and strains do not matter too much, but in the larger sheets needed for photography it is critical. It is also essential to check that the double reflections do not interfere. These double reflections tend to be more of a problem with mirror optics than ordinary glass lenses. It is necessary to check the optical properties thoroughly, and partial eclipses and regular solar observations present good opportunities for doing this.

The sheets are considerably heavier than most other filters, and the sheer weight of them can unbalance lightweight optical equipment such as mirror lenses when put on and removed. The image colour varies from neutral to a bilious green. Fortunately, the colour bias is unimportant in monochrome, and is well within the scope of correction in normal photographic and video reproduction.

Choose welding glass where safety is paramount – genuine suppliers apply internationally agreed standards and supply certificates. Welding glass is also very difficult to break, because of its main application under hazardous conditions. Its thickness is another advantage as it still works well enough for our less critical purposes even when scratched! The ruggedness has to make welding glass a good choice for all the less critical optical applications.

Consider these filters for binoculars used in routine scanning and monitoring. Reserve the top quality filters for any more precise work through the main optics. Welding glass can be Araldited into a rigid mount, which can then be securely fixed to the front of the binocular optics.

Wratten No. 96 Filters

These filters used to be made from gelatine, and were extremely difficult to keep in top condition. The modern plastic material is considerably more robust but still prone to scratches. Only consider pristine filters for the main optics. Because of their origin in illumination, rather than photography, they had to be transparent to heat, i.e. IR. If they were not, they burst into flames! Consequently, Wratten filters of all sorts, particularly neutral density and No. 96, must never be used as the sole filter whenever the image is viewed by the naked eye. Too much IR is transmitted to be safe. The heat energy can also damage delicate iris mechanisms at any focus. Better to avoid the filter type altogether where there is any risk to either the eye or the optical system.

Never use a Wratten filter with the modern video and CCD technologies. The detectors are extremely sensitive in the IR. The scene might appear safe enough behind a Wratten 96 as no visual image is seen, but the full blast of solar IR heating is still going through. Many of the modern pieces of equipment have safety cutouts and heat filters for just such an accident as pointing the lens directly at the Sun. It is simply best to forget Wrattens for video/CCD work, just in case you blow the electronics by mistake.

However, Wrattens are probably the best choice for preset or rangefinder cameras, i.e. where the image is not seen directly. The main optics are well protected in routine usage, where the camera can be carried around the neck all day, exposed to full sunlight. Simply run a series of test exposures, using the wide range of densities offered, and work through the camera speeds, and stops if offered. Make full notes, and the system is calibrated for all time.

Filters Which Must Never be Used

Smoked Glass

Smoked glass has a venerable history, and was the only filter available in days of yore. Huge numbers were made for use in schools etc., and that is the reason why they must not be used now – there is no quality control, and there have been too many accidents in recent years to take any more chances. The black deposit from a candle or smoky flame is a horrible mixture; it is not just carbon black, but contains unburnt wax and oil. It is almost transparent to IR and is quite unsuitable as a solar filter for any optical instrument.

Smoked glass is simply too dangerous for its original main purpose, which was to hold up for direct views of the Sun with the naked eye. Cases are on record of breakages, with obvious puncture injuries, but the smoke deposit is uneven by its very nature. Any pinprick already present or caused by fingering gives an almost perfect pinhole camera to focus damaging IR in the eye. Remember, accidents have happened. Don't use smoked glass.

"Black" Colour Film

All colour films lack silver and are just dyed images, no matter how dense they seem. Full IR is allowed through, making them very dangerous indeed.

Silver Printed Plastics

These have been referred to as "glitters" when discussing Mylar. No matter how suitable they might appear to be, the risk from using the wrong stuff is so high that it is worthwhile repeating the prohibition again.

Polaroids

Crossed polaroids will give an incredibly dense filter which is fine for ordinary photography. Most polaroid filters are based on the same sort of technology used in Wratten filters and are designed to let IR radiation through. It is much easier and safer to think of all polaroids in this way than to pick up the wrong sort by mistake. There is also the question of multiple reflections from the four surfaces.

Prisms

Special direct-view solar viewers used to be on sale. They went under a variety of names, and "helioscope" was one of many. The designs were based on multiple reflection off prisms, such as a Kolzi-prism, very much like binoculars, except that the prisms were unsilvered. All old equipment must be regarded as suspect until checked out by an expert.

Hydrogen-Alpha Filters

An H-alpha filter is simply a specialised normal filter designed to pass a very narrow wavelength around the C line in the Fraunhöfer spectrum. This is centred about 656 nm, in the red. These filters have revolutionised solar studies, and have largely superseded the complex and specialised spectroheliographs and most spectrohelioscopes. The single filter replaces several prisms or gratings, and no irksome collimation is needed.

One of the earliest H-alpha filters was devised by Bernard Lyot to go with his coronagraph. It was not a true H-alpha filter as the most important line in the coronal spectrum at the time was the green line. At that wavelength it achieved an amazing transmission only 0.3 nm wide. A special extra device was needed to home in on the next important line of the time, H-alpha. Photographic plates of the early 1930s were also at their most sensitive in the blue to green spectrum and were notoriously insensitive in the red. The more interesting H-alpha prominences in the outer regions of the corona are very much fainter than the inner chromosphere, some hundred to a thousand times. It was not until 1948 that Lyot was able to perfect his technology to allow direct observations in these outer regions. His new filters had a bandwidth a fifth of the earlier ones.

A few more years passed before Rosch and Dollfus were able to achieve filters capable of imaging the whole Sun rather than the rim. This opened up a whole new research field since it was now possible to study the relationship between the more familiar prominences and the various surface phenomena.

In the 1970s production of interference filters had become sufficiently standardised that they were affordable in the amateur market. The early varieties worked extremely well, but had their own complications in needing a lot of power away from a fixed observatory during eclipse expeditions.

The filter only worked when heated and the temperature needed to be controlled to a narrow range. Changes in temperature altered the transmitted wavelength, and this was how velocity and depth measurements were made. Modern H-alpha filters still need very careful handling, and are best dedicated to fixed observatories where a sensible observing programme can be carried out over a long period.

There has been an exciting development in recent years as a spin-off from the light pollution problem. Fixed bandwidth H-alpha filters have become available in large sizes which are just like any other glass filter. Their original intention is for normal astronomy at night to pick up H-alpha emission regions and nebulosity. These are quite difficult to pick up at the best of times, and virtually impossible in light-polluted areas. The average human eye is not too sensitive in the red, and about a third of men suffer from some sort of red deficiency, making it doubly difficult. With one of these filters, the red regions are all you see against a black background. Photography is also that much simpler.

The filters come in quite large sizes and are placed over the front of the primary optics, such as a telephoto lens (see Fig. 3.4). Some companies also offer varieties for monochrome or colour film. Whilst not suitable for use on their own for solar work, they can be combined with a standard Inconel filter for a rapid survey of solar activity and potential prominences at eclipse time.

Some other very stable filters have been made which reject all but a narrow wavelength. These "light rejection" filters are usually quite small, and because they are fitted at the eyepieces are totally unsuitable on their own for solar work. However, combined with a proper filter (e.g. Inconel) over the main optics, a much wider range of solar wavelengths can be studied at modest cost. These modern filters can be described as the

Figure 3.4. Eric Strach using an H-alpha filter in Kenya, 1980.

descendants of the pioneering spectrohelioscopes and spectroheliographs. Combining sensitivity of CCD and similar video technology, a new breed of coronagraphs can also be expected to reach the amateur market. Detailed studies of the inner chromosphere are still best carried out with some variety of coronagraph, where the bright disk of the Sun can be eliminated entirely.

Spectroheliographs and Spectrohelioscopes

Some confusion might be found with the terminology of this topic. The principles of viewing or recording the Sun's radiation at a desired wavelength are simple. It is not so obvious that great care has to be paid in interpreting the results.

All the interesting solar spectral lines are actually *absorption* lines, which means that we are only dealing with residual light left over in a much brighter background. When spectra of the other stars are taken, this does not matter so much, as the lines appear dark or completely blank. They are an integration for the whole disk. For the Sun the problems are greater since we need to record the much fainter radiation within a tiny area, in its much brighter surroundings, which in turn is submerged in the brightness of the whole spectrum, including heat.

The spectral lines are brightly emitted light but at much reduced radiation. In everyday terms, if the lines were seen in isolation they would outshine anything man-made in the same way as sunspots, but against the Sun's brighter lighting, it is an achievement to record anything at all. The prominences are regions of light emission, which are very faint relative to the whole corona. They can only be seen to perfection with the naked eye in the extended regions away from the Sun's limb at times of total eclipse. Prominences glow with the same H-alpha radiation seen in extended gas clouds in nebulae such as the Trifid Nebula, M20 in Sagittarius.

Strictly speaking, a spectrohelioscope is simply a device for looking at the Sun's spectrum at any desired wavelength, and is nothing more than an ordinary spectroscope. The major design difference is the need to get rid of the bulk of the heat and other undesirable energies, whereas in ordinary stellar spectroscopy we have the totally different problem of not enough energy. A spectroheliograph, on the other hand, is an instrument where a permanent record of the spectrum is made. Until fairly recently, the record was only possible on black and white film, then colour, but that has all changed with the advent of electronic technologies. Many of the older instruments are still available, and still work very well. It does not matter what form of permanent record is used, it is still called a spectroheliogram, be it a silver-halide film, video, or computer print-out.

All forms of solar spectrum study used to rely heavily on prisms or gratings to make the necessary wavelength selection. Each type had its own advantages and disadvantages, and either is still the way to go when construction cost is the major concern. A typical spectroheliograph will have three or more prisms in it, and the normal design will arrange these in a separate box so that the whole can be used as a movable monochromator.

The recent advent of cheaper and much larger interference filters has dramatically altered the situation for professionals and anyone needing more precise measurements. Each filter can be tailored to any wavelength of interest, and the optical collimation is much simplified at the same time. Some fascinating spectrohelioscope designs have appeared over the years, although all have the same basic features. Because the Sun is scanned in only a small slit it is not possible to get a complete picture of the whole disk at our chosen wavelength. However, some workers are only interested in the prominences in H-alpha at the Sun's limb, so that this system works excellently for them, and a simple spectrohelioscope will do.

A variant of the basic design was invented in 1930 by B. Lyot in France. He produced an artificial eclipse in the light path in the form of a small cone which reflected all the Sun's light. This is the coronagraph which revolutionised solar studies. It allows views and pictures of the prominences any day, not just at total eclipses. It requires extreme care in construction to eliminate stray light problems and is rarely found outside professional observatories.

When it is necessary to see all the Sun's disk in H-alpha or some other wavelength, some more elaborate technology has to be brought into play. Because the eye cannot take in the whole of the Sun's disk through a slit, the slit is made to oscillate to and fro. The well-known phenomenon of persistence of vision then takes over, just as in a cinema, and the picture becomes clear for the whole area.

A well-designed spectrohelioscope will have a device built into it to select wavelengths close to the normal one so that areas moving at high speeds can be allowed for and measured accurately. Such devices for velocity measurement are prosaically referred to as "line-shifters" which amply sums up their function. In older designs the line shifter had to be some form of mechanical device, whereas the modern interference filters can be controlled electronically in any mode.

A spectroheliograph, on the other hand, can be just a simple visual instrument with a facility for substituting the recorder when needed. Most of them are designed specially with direct naked-eye observational accessories present only for monitoring and to check all is in order. Some of the professional equipment is impressive, and on a grand scale. Very few of these establishments rely on a direct tracking of the Sun with a telescope. The Sun is directed into specialised and rigid equipment underground by a system of mirrors referred to as a heliostat, or sometimes a single mirror called a siderostat or coelostat. The main disadvantage of the single mirror option is that the Sun's image rotates through its centre during the observation.

Perhaps the most impressive solar telescope is the McMath instrument at Kitt Peak (see Fig. 3.5), which has a primary heliostat two metres in diameter.

The solar tower at Mount Wilson at 45 metres tall is another familiar sight (see Fig. 3.6).

Some dedicated amateurs such as Commander Henry Hatfield have built their houses around their spectro-helioscopes.

What we very quickly find out is that the visual or recorded picture is not always as "clear" as we would like, with either the interesting patches disappearing or missing altogether, or changing noticeably in a period of a few minutes. This is quite normal, and due entirely to the enormous speeds at which some of the gas clouds, faculae and prominences are moving, in turn due to the well-known Doppler effect. A well-designed instrument will have a facility to select wavelengths close to the normal one, so that these speeds can be allowed for and measured accurately.

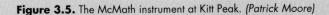

Figure 3.5. The McMath instrument at Kitt Peak. *(Patrick Moore)*

Figure 3.6. The solar tower at Mount Wilson. (Patrick Moore)

Because the absorption lines are seen through the solar atmosphere, and the absorption varies with depth, some very useful work can be done in measuring the depths as well as the velocities and magnetic fields. The specialised instruments for measuring magnetic fields, the Zeeman effects, are called magnetographs and go back to the pioneering work of H.D. and H.W. Babcock. Even more specialised equipment is used to measure the radio spectrum using interferometers. These radio-heliograms are outside our present scope.

Observing Sunspots

If it is so unsafe to use a telescope for direct solar observation, then how can we observe sunspots?

The easiest and the most basic method is to use the telescope as a projector; for this purpose a refractor is much more convenient than a reflector, though of course reflectors can also be used. With a relatively large telescope, it is always worth stopping down the aperture; after all, with the Sun there is always plenty of light to spare.

First, line the telescope up with the Sun, taking care to keep your eye well away from the eyepiece. The photograph here (Fig. 3.7) shows a five-inch refractor fitted up for solar work. The telescope is aimed sunward, and the shadows cast on the back of the screen show that the alignment is correct, so that the Sun's image will pass through the main telescope on to the screen held behind the eyepiece, and any spots which happen to be on the disk will show up at once. The size of the projected disk depends partly upon the magnification being used, and partly on the distance between the eyepiece and the screen; the greater the separation, the larger the disk. All in all, a six-inch disk is a good compromise, though every observer will have his own personal preference. The next step will be to make a projection box; if the telescope is mechanically driven (or even if it is not), this makes sunspot drawing a great deal easier.

Figure 3.7. Patrick Moore shows how to use a simple hand-held screen for solar projection.

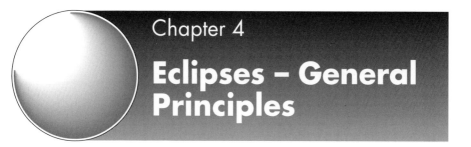

Chapter 4
Eclipses – General Principles

Lunar and solar eclipses are completely different. A lunar eclipse occurs when the Moon passes into the cone of shadow cast by the Earth; since the Moon shines only by reflected sunlight, it turns a dim, often coppery colour when the supply of direct sunlight is cut off. A solar eclipse takes place when the Moon passes in front of the Sun, blocking it out for a few precious moments. Strictly speaking, a solar eclipse is not an eclipse at all, but an occultation of the Sun by the Moon.[*]

Lunar Eclipses

As a start, let us look at lunar eclipses, albeit briefly. First, it may be asked why the Moon does not vanish completely when it passes into the shadow. The reason is that some of the Sun's rays are bent or refracted on to the lunar surface by way of the shell of atmosphere surrounding the Earth, and it follows that the darkness of the eclipse depends upon conditions in our own

---

[*] Occultations of stars by the Moon are common enough, though not many really bright stars are sufficiently close to the ecliptic to be occulted; of the first-magnitude stars only Alderbaran, Antares, Regulus and Spica qualify. Planets may also be occulted by the Moon, and occasionally a planet will occult a star; occultations of one planet by another are very rare. Stars and planets can be occulted by the Sun, but these phenomena are obviously quite unobservable!

upper air. If the higher layers of the atmosphere are loaded with dust or ash – as happens, for example, after a major volcanic eruption – the eclipse will be dark; it is said that in 1816, after a tremendous eruption in the East Indies, the Moon could not be seen at all during totality. On the other hand, the eclipse of 1848 was so bright that lay observers refused to believe that an eclipse was happening at all. It is worth noting that the recent eruption of Mount Pinatubo, in the Philippines, produced several eclipses which were unusually dark.

The French astronomer A. Danjon once drew up an eclipse scale, from 0 (dark) to 4 (bright); he also tried, without much success, to correlate eclipse brightness with sunspot activity. The Danjon scale is as follows:

0:   Very dark. Moon almost invisible.
1:   Dark grey or brownish colour; details barely identifiable.
2:   Dark or rusty red, with a dark patch in the middle of the shadow, and brighter edges.
3:   Brick red, sometimes with a bright yellow border to the shadow.
4:   Coppery or orange-red, and very bright, sometimes with a bluish cast and varied hue.

Not all eclipses are total; very often only part of the Moon passes into the shadow. Before entering the main umbra, or cone of shadow, the Moon must pass through the "penumbra", or zone of partial shadow, which exists because the Sun appears as a disk rather than a source. There are also eclipses in which the Moon never enters the umbra at all. On these occasions the penumbra is detectable without much difficulty as a slight but appreciable dimming of the Moon's surface.

At the mean distance of the Moon, the diameter of the Earth's shadow cone is approximately 5700 miles (9170 km), and the average shadow length is 850,000 miles (1,367,650 km). Totality may last for as long as 1 hour 44 minutes, so that a lunar eclipse is a comparatively leisurely affair. Needless to say, a lunar eclipse can happen only at the time of the full moon, and it can be seen from anywhere on Earth where the Moon is above the horizon at the time; this is why as seen from any one location, eclipses of the Moon are more common than those of the Sun. They do not happen at every full moon, because the lunar orbit is tilted at an angle of just over 5 degrees.

It cannot honestly be said that lunar eclipses are of real astronomical importance. It is true that as the shadow sweeps across the Moon there is a sudden drop in the Moon's surface temperature, and it was once believed that this might produce detectable effects in some formations, but this does not seem to be true – though there are various regions, known misleadingly as "hot spots", where the temperature fall is less marked than elsewhere, because of differences in the surface structure or composition (the best example being the great ray crater Tycho, in the Moon's southern uplands). But at least lunar eclipses are lovely to watch, and are great favourites with astrophotographers.

For the sake of completeness, tables of lunar eclipses for the period 1998–2020 are given in Chapter 13. But since this book is about eclipses of the Sun, not those of the Moon, let us turn to our main theme.

Solar Eclipse Types

Solar eclipses can happen only at a new moon, and again they do not happen every month because of the tilt of the lunar orbit – at most new moons, the Moon passes either above or below the Sun in the sky, and there is no eclipse. By a fortunate chance, the Sun and the Moon appear almost the same size in the sky; the Sun's diameter is 400 times that of the Moon, but it is also nearly 400 times further away from us. If the Moon appeared slightly smaller than it actually does, it could never produce totality; if it looked slightly larger, it would cover not only the Sun's photosphere but also the chromosphere, prominences and inner corona, so that the full beauty of the spectacle would be lost.

Solar eclipses are of three types:

1. *Total.* The Sun's photosphere is completely hidden. While it is wholly blacked out, the solar atmosphere flashes into view, and the sight is magnificent. But since the Moon's shadow is only just long enough to reach the Earth, totality is brief; nowhere can it last for more than 7 minutes 31 seconds, and most eclipses are much shorter than this (sometimes only a second or two). Moreover, the track of totality is narrow, and cannot be more than 169 miles (272 km) wide, so that

the would-be observer has to be in exactly the right place at exactly the right time.

2. *Partial.* The Sun's disk is only partially hidden by the Moon. A partial eclipse is seen from regions to either side of the belt of totality, but there are also partial eclipses which are not total anywhere on the Earth. Moreover, the slightest sliver of photosphere remaining uncovered is sufficient to hide the prominences, chromosphere and corona, so that a partial eclipse is unspectacular and of relatively little interest.

3. *Annular.* The Moon's orbit round the Earth is not circular, and neither is the orbit of the Earth round the Sun, so that the apparent diameters of both Sun and Moon vary. The limits are as follows:

Sun: max. 32′ 35″	mean	32′ 01″	min. 31′ 31″
Moon: 33′ 31″		31′ 05″	29′ 22″

At some central alignments, therefore, the Moon's shadow is too short to reach the Earth, and at mid-phase a ring of sunlight is left showing round the dark disk of the Moon – hence the name (annulus is Latin for "ring"). Again, the Sun's surroundings cannot be properly seen, though at some annular eclipses there are interesting phenomena to be observed – and are discussed in Chapter 8. The longest possible duration of annularity is 12 minutes 24 seconds, and again a partial eclipse is seen from regions to either side of the central lines.

The length of the Moon's shadow ranges between 237,000 miles (381,000 km) and 227,000 miles (365,000 km), with a mean of 231,000 miles (372,000 km). As the mean distance of the Moon from the Earth is 238,600 miles (384,000 km), the shadow is on average too short to reach the surface, so that annular eclipses are more frequent than totalities in the ratio of 5 to 4. On average there are 238 total eclipses per century, but from any particular location they are depressingly rare. For example, from England there were only two total eclipses during the twentieth century – those of 1927 and 1999 – and even these were seen from restricted areas. English observers must then wait until 2081, when the track of totality will brush the tip of Cornwall. Today, with space research, the situation has been transformed, but total eclipses are still astronomically important.

Quite apart from this, the beauty of the spectacle is unrivalled. It is fair to say that there is nothing in the whole of Nature to rival the glory of a total eclipse of the Sun.

The Sequence of Events

In 1965, one of the authors (PM) made his way to Siberia, in order to observe a total eclipse. The chosen site was littered with equipment of all kinds, but there was one anomaly. A famous Dutch astronomer, who had made notable contributions to solar physics, greeted us cheerfully; his equipment seemed to consist solely of an armchair. Questioned, he willingly explained. "I have been to many eclipses," he said. "I have made studies of many eclipses. Never have I been able to watch an eclipse. So this time – I do absolutely nothing!" And he sat there throughout totality, enjoying the scene and making no attempt to carry out work of any kind.

In this he was eminently sensible, and there is a moral here. When you go to an eclipse, make scientific observations by all means; but do not forget to take a certain amount of "time off" to appreciate the spectacle which Nature is providing.

Let us, then, run through some of the different phenomena that will be observed during a total eclipse, more or less in the order in which they occur.

First Contact. A tiny notch appears in the Sun's limb, heralding the advance of the Moon. Gradually the notch increases, and before long it becomes very evident, though there will be no noticeable diminution in sunlight or heat until at least half the disk is covered. By all means take photographs, but be wary of spending too much film on the pre-totality partial phase, much better to do this after totality, when you know how much film you have left! When the sky begins to lose its brightness, look for Venus or any other bright planet which happens to be favourably placed.

Crescent Images. When the visible Sun has been reduced to a crescent, look for crescent images on the ground if you are standing anywhere near a tree. The spaces between the tree-leaves act rather in the manner of pinhole cameras, and the effect is very marked and well worth photographing (see Fig. 3.1).

Baily's Beads. These should really be known as "Halley's Beads", since it was Edmond Halley who first recorded them during the eclipse of 1715, but it was Francis Baily who gave the first really detailed description of them, in 1836 – actually this was an annular eclipse, but the Beads are better seen just before totality. Let us quote Baily's own description:

> When the cusps of the Sun were about 40° asunder, a row of lucid points, like a string of bright beads, irregular in size and distance from each other, suddenly formed round that part of the circumference of the Moon that was about to enter on the Sun's disk. Its formation, indeed, was so rapid that it presented the appearance of having been caused by the ignition of a fine train of gunpowder. Finally, as the Moon pursued her course, the dark intervening spaces (which at their origin, had the appearance of lunar mountains in high relief, and which still continued attached to the Sun's border), were stretched out into long, black, thick parallel lines, joining the limbs of the Sun and Moon; when all at once they suddenly gave way, and left the circumference of the Sun and Moon in those points, as in the rest, comparatively smooth and circular, and the Moon perceptibly advanced on the face of the Sun.

There is no mystery about the Beads; the lunar surface is uneven, so that the sunlight shines through the valleys and is blocked by the mountains. Obviously their appearance depends upon the particular circumstances of the eclipse, but they are always seen.

The Diamond Ring. Just before Second Contact, the last segment of the Sun shines out against the lunar limb, and the effect is glorious; it is very easy to see why it is always termed the Diamond Ring. It does not last for more than a few seconds; almost at once it vanishes, and the corona comes into view.

Second Contact. Totality! It happens with amazing suddenness; the Diamond Ring and Baily's Beads are gone, and the sky has darkened.

The degree of darkness is variable, mainly because of conditions in our own air; sometimes the stars shine out brilliantly, while at other eclipses they cannot be seen at all; though any conveniently-placed planets will generally come into view provided that there is no cloud or haze.

Chromosphere and Prominences. The red "colour-sphere" is striking as soon as totality begins, and there may well be one or more prominences; not unnaturally, the prominences were originally known as "red flames". Let us quote Francis Baily again:

> [The prominences] had the appearance of mountains of a prodigious elevation; their colour was red, tinged with lilac or purple; perhaps the colour of the peach-blossom would more nearly represent it. They somewhat resembled the snowy drops of the Alpine mountains when coloured by the rising or setting sun. They resembled the Alpine mountains also in another respect, inasmuch as their light was perfectly steady, and had none of that flickering or sparkling motion so visible on other parts of the corona. They were visible even to the last moment of total obscuration; and when the first ray of light was admitted from the Sun, they vanished, with the corona, altogether, and daylight was instantaneously restored.

Shadow Bands. In 1820 the German astronomer Hermann Goldschmidt was the first to notice wavy lines seen across the Earth's surface just before totality. These so-called shadow bands shift quickly, and are not always seen; they are best displayed against bright surfaces such as whitewashed walls. They are, of course, purely atmospheric phenomena – and are surprisingly difficult to capture on film.

Shadow of the Moon. As totality approaches, the Moon's shadow can be seen sweeping across the landscape – or, even better, the seascape. It travels at over 1000 m.p.h., and gives the impression of a vast dark cloak rushing towards you and then enveloping you: the general effect can only be described as eerie. It is not easy to photograph the oncoming shadow, and almost before you have time to appreciate it you find that totality has begun.

The Corona. The wonderful "pearly mist" is unquestionably the most beautiful of all the phenomena of totality. Probably the first mention of it is due to the Roman writer Plutarch, who lived from about AD 40 to 120; he refers to "a certain splendour" round the eclipsed sun which must surely have been the corona. It was also described by observers in Corfu during the eclipse of 22 December 968. Christopher Clavius saw it on 9 April 1567, but made the mistake of assuming that it was merely an effect of the uncovered edge of the

Sun. Johannes Kepler believed that it was due to a lunar atmosphere, and this remained the general view for some time, though at the eclipse of 16 June 1806 the Spanish astronomer Don José Joaquin de Ferrer pointed out that in this case the Moon's atmosphere would have to be at least fifty times the extent of that of the Earth, which did not seem very likely. The question was finally settled during the eclipses of 1842 and 1851, when it was seen that the Moon's disk passed steadily across the corona and must therefore be placed in the foreground.

During totality, the corona radiates as much light as a normal full moon, and this means that direct observation is safe – though never forget that the total phase ends very suddenly. The shape of the corona varies according to the state of the solar cycle; near spot-minimum it is more or less symmetrical, while when the Sun is active, with many spot-groups, there are long coronal streamers which looks remarkably like wings. During the Maunder Minimum of 1745 to 1815 there is some evidence that the corona was very inconspicuous, though unfortunately the records are far from complete.

We have already referred to the Maunder Minimum. (To be scrupulously fair it should be called the Schwabe–Maunder Minimum, because it was Heinrich Schwabe who first identified it, though it was admittedly E.W. Maunder who publicised it.) The records for the 1645–1715 period are incomplete, but it does seem that there was a dearth of sunspots, and certainly this coincided with what has been called the "Little Ice Age"; in England, during the 1680s the Thames froze every year, and frost fairs were held on it. Whether or not the cold spell has any connection with the quietness of the solar disk is by no means established, but certainly sunspots have a profound influence on the phenomena of totality, and it may well be that during the Minimum the corona was very inconspicuous. It is a great pity that we do not have detailed records.

Photographing the corona is described in Chapter 10. In the pre-photographic age the best records were obtained by drawing, and some of the old sketches are fascinating; one, by the Belgian astronomer Etienne Trouvelot, is reproduced in Fig. 4.1.

This was at the eclipse of 6 May 1883, which Trouvelot saw from Caroline Island in the South Seas. He also made some comments about the island itself – which was inhabited by four men, one woman and three children, whose lives were "absolutely eventless"!

Figure 4.1.
Trouvelot's sketch of the 1883 eclipse.

Third Contact. Totality is over. Once more there is a brief glimpse of the Diamond Ring: the brilliant speck swells, and smears out the corona, chromosphere and prominences as sunlight returns to the world. Nature awakes with almost incredible suddenness, and in only a few seconds it almost seems that the eclipse has never been.

As the Moon draws away, this is the time to take photographs of the partial phase; to note again the crescent images cast by tree leaves, and to record the time when brilliant planets such as Venus and Jupiter disappear.

Fourth Contact. The last fraction of the lunar disk leaves the Sun.

No two eclipses are alike, and one never knows quite what to expect. So be prepared for any eventuality. Above all, work out your programme well in advance, and rehearse it so thoroughly that you are absolutely certain of what you intend to do during the fleeting moments of totality.

Chapter 5

Eclipses in History

Eclipses have been recorded since very early times – which is not at all surprising; a total solar eclipse, at least, could hardly be overlooked! Stories about them are many and varied, so let us begin with one which, although certainly of doubtful origin, has been widely told.

It dates back to the year BC 2136, in the reign of the Chinese emperor Chung K'ang. At that time the Chinese had no idea of the cause of eclipses, and put them down to the attacks of a hungry dragon which was doing its best to gobble up the Sun. The remedy, of course, was to beat gongs and drums, shout and scream, and make as much noise as possible in order to scare the brute away. The procedure invariably worked, but it was always necessary to be on the alert, and this was the responsibility of the Court Astrologers, who at this time rejoiced in the names of Hi and Ho.

Judge the consternation, therefore, when it was suddenly realised that the dragon was once more on the attack! No warning had been given, and although the eclipse ended after a few hours the Emperor was far from pleased. In fact he ordered the immediate execution of the two culprits, leading on to a rhyme whose authorship is unknown:

> Here lie the bodies of Ho and Hi,
> Whose fate, though sad, is risible;
> Being slain because they could not spy
> Th' eclipse which was invisible.

Unfortunately, this episode has been discounted by modern scholars. However, it is an intriguing story.

Let us turn next to the eclipse of 28 May BC 585, which is said to have been predicted by Thales, first of the great Greek philosophers. At least we are fairly sure of Thales' own dates – he lived from around BC 624 to 547 – and the eclipse certainly happened; according to legend it interrupted a battle between the forces of King Alyattes of the Lydians and King Cyaxares of the Medes who were so alarmed that they called a hasty truce. But could Thales, wise man though he undoubtedly was, have predicted the eclipse at all?

He did not know the cause; he believed that the Earth was flat, and floated on a vast ocean. If he really did make the predication, there is only one way in which he could have done so: by using the Saros.

The interval between one new moon and the next (or one full moon and the next) is termed a lunation; its length is 29.53 days, or, to be more precise, 29 days 12 hours 44 minutes 2.9 seconds. After 223 lunations, amounting to 18 years 11 days, the Earth, Sun and Moon return to almost the same relative positions; this 18-year period is called a Saros. It means that any particular eclipse is likely to be followed by another eclipse 18 years 11 days later. For example, consider the eclipses of two Saros series:

July 9 1945, July 20 1963, July 31 1981, August 11 1999, August 21 2017; and
May 29 1919, June 8 1937, June 20 1955, June 30 1973, July 11 1991, July 22 2009

and so on. The relationship is very marked, and there are many Saros series running at the same time. The first eclipse of any particular cycle begins as a partial, near one or other of the Earth's poles; successive eclipses reach lower and lower latitudes and in most cases become total; then, near the end of the cycle, there are final partial eclipses near the Earth's opposite pole. The Saros which included the 1991 eclipse began on 22 June 1360, and will end on 30 July 2622; it will have contained 71 eclipses in all, of which 45 will have been total.

The relationship is not exact, so that successive eclipses in a Saros cycle are not identical; for example the 1995 eclipse was total over India, but the eclipse of 2013 is annular/total. All the same, the Saros is a reasonably reliable guide, so could Thales have know about it?

Frankly, we cannot be sure, and neither can we be really confident that he predicted the eclipse of BC 585, but it is a distinct possibility. Certainly the later Greek philosophers knew about the Saros, and they also knew that a solar eclipse was due to the blotting-out of the Sun by the Moon. Hungry dragons were quite definitely ruled out, even though the Japanese made a habit of covering their wells during eclipses to prevent poison being dropped into them from the sky!

The Peloponnesian War, between the Greek city-states of Athens and Sparta is associated with two eclipses. Undoubtedly this was one of the most important conflicts of classical times; in the end the Athenians were beaten by the military might of Sparta, but it might so easily have gone the other way, in which case the whole of Mediterranean history would have been different. In the first year of the war, BC 431, there was an annular eclipse just as Pericles, the Athenian commander, was about to set sail with his fleet. The sailors were alarmed – but Pericles reassured them by holding a cloak in front of his eyes, and assuring the sailors that the Sun was merely being hidden by something larger than a cloak.

The second Peloponnesian eclipse was in fact a lunar one. It fell on 27 August BC 413, when the Athenians had been unwise enough to send a military expedition to Sicily and had found themselves in deep trouble. They could have escaped by sea, but the official astrologer warned Nicias, the Athenian commander, that because of an impending eclipse of the Moon he must stay in port "for thrice nine days". Nicias agreed. Nothing could have suited the Spartans better; when Nicias finally decided upon evacuation, he found that he had been very effectively blockaded. His fleet was wiped out, and so was the entire Athenian force. Only a few years later Athens was forced to surrender, and the city's famous Long Walls were pulled down.

Another naval commander – Agathocles, tyrant of Syracuse – was associated with an eclipse in BC 310. He had been blockaded in Syracuse by the Carthaginian fleet, but managed to escape one night while the enemy vessels were attacking a provision convoy. The Carthaginians gave chase, but when the eclipse began the sudden darkness caused such confusion that Agathocles was able to slip away – after which he proceeded to land in Africa and plunder Carthaginian territory.

There are vague reports of unusual darkness at the time of the death of Julius Caesar, and again at the Crucifixion of Christ, but neither of these seems to be well authenticated, and certainly cannot be linked to any eclipses. However, there was certainly a solar eclipse in August AD 45 which happened to be the birthday of the Roman emperor Claudius. Historians will remember Claudius as a reluctant ruler who was literally pitchforked into the role after the murder of Caligula, who had caused major crises because he was completely mad (he even elevated his horse to the status of consul). To everybody's astonishment, Claudius proved to be an extremely capable ruler,* and he was also a scholar. To quote the Roman writer Dion Cassius: "As there was going to be an eclipse on his birthday, through fear of a disturbance, as there had been other prodigies, he put forth a public notice, not only that the obscuration would take place, and about the time and magnitude of it, but also about the causes that produce such an event."

In AD 453, when Attila the Hun was raiding Italy, the writer Gregorius Turonensis records that on 24 February "even the Sun appeared hideous, so that scarcely a third part of it gave light, I believe on account of such deeds of wickedness and the shedding of innocent blood." In 716 we find the following entry in the *Anglo-Saxon Chronicle*: "In this year Ethelbald captured Somerton, and the Sun was eclipsed, and all the Sun's disk was like a black shield, and Acca was driven from his bishopric." This seems to have been an annular eclipse; what happened to the luckless Acca is not stated.

Another major battle took place in 1030 at Sticklestead, not far from Trondheim in modern Norway. The forces concerned were those of the Norwegian king Olaf (an early convert to Christianity) and King Canute, who ruled England as well as Denmark. The battle was described by the Icelandic poet Sighvald:

> They call it a great wonder
> That the Sun would not
> though the sky was cloudless
> Shine warm upon the men

Whether the battle actually took place on the day of the eclipse is not certain, but it may well have done. At any rate, King Olaf was killed. Earlier, in 840, the emperor Louis of Bavaria is said to have been so

---

* *En passant*: it is worth remembering that it was Claudius who managed to conquer Britain; Julius Caesar had merely carried out a lightning raid.

terrified by an eclipse that he collapsed and died – after which his three sons proceeded to indulge in a ruinous war over the succession.

Coming forward to 1346, and the Battle of Creçy between the French and the English, we find the following account given in a famous book, *History of England*, by Lingard: "Never, perhaps, were preparations for battle made under conditions so truly awful. On that very day the Sun suffered a partial eclipse; birds in clouds, the precursors of a storm, flew screaming over the two armies, and the rain fell in torrents, accompanied by incessant thunder and lightning." This sounds very convincing, but in fact we are quite certain about the date of the Battle of Creçy, and there was no eclipse visible from Europe anywhere near that time.

Gradually the superstitious fear of eclipses lessened, but at least the old records are scientifically valuable because they can help us to measure what is termed secular acceleration.

Most people know that the Moon spins on its axis in exactly the same time that it takes to complete one orbit, so that the same face is always turned toward the Earth; before the circum-lunar flight of the Russian space-craft *Lunik 3*, in 1959, we had no direct information about the far side of the Moon. This is not mere coincidence; it is due to tidal effects. Initially the Moon was much closer to the Earth than it is now, and the two bodies raised tides in each other, bearing in mind that both were viscous rather than solid. Since the Earth is 81 times as massive as the Moon, the tides in the lunar globe were very strong. As the Moon rotated, the Earth tried to keep a "bulge" pointing in our direction, and the Moon's spin was slowed down until relative to the Earth – though not relative to the Sun – it had stopped altogether. All the other large planetary satellites in the Solar System have similarly captured or synchronous rotation.

Similarly, the Moon's tidal pull is braking the rotation of the Earth, so that the days are becoming longer – not by much; a tiny fraction of a second per century, but over a sufficient period of time the effect shows up, even though there are irregularities due to movements inside the Earth itself. A second result of this tidal braking is that the Moon's distance from the Earth is increasing at a rate of between 4 and 5 centimetres per year.

These secular effects mean that if we calculate the times of ancient eclipses, we will find a slight, systematic error. Obviously the old records are by no means precise, but they are good enough for us to be quite confident of the effect.

When will Earth see its last total eclipse? Well, it seems that in about 1,250,000 years hence the Moon will have moved outward by an extra 18,000 miles or so, and its disk will no longer be large enough to cover the Sun. If men still inhabit the world at that remote epoch, they will no longer be able to enjoy the beauty of a total solar eclipse, but at least any readers of the present book have the consolation of knowing that in our own age, and in the foreseeable future, total eclipses will continue.

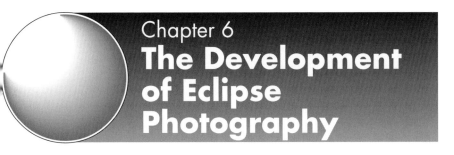

Chapter 6
The Development of Eclipse Photography

A moment's thought and it is obvious that eclipses have influenced science as a whole. No scientific development is more important to astronomers than photography, which has had such a profound influence in return on the type and accuracy of all scientific measurements. Only by making a digression into the history of photography can we begin to have an understanding of why the Sun, and particularly the Sun in eclipse, is so pivotal to so much of the science we take for granted.

Photographing the Sun is easy – millions of people do it every day. Without the Sun there would be no light and cameras and camcorders would not work. Although that application is not strictly what we mean by solar photography, it is a start. An historical review also gives very strong clues on the way to go about our chosen subject and will hint at, and perhaps answer, the many problems we may face.

Photograms

The first problem, painfully obvious in every sense of the word, is sunlight's sheer power. As we shall see, many problems arise from the different actinic energies in the visible and invisible parts of the spectrum, and their optical properties in imaging systems.

Long before mankind understood the nature of sunlight, many ideas were developed to harness solar power to record an image. Perhaps the most obvious is sunlight's ability to bleach dyes and other coloured materials. Any area protected from the Sun retained its original colour, which showed up clearly on the exposed and bleached background. A little ingenuity and the image could be reversed by using a stencil. The principle is still with us today in a whole branch of photography called photograms.

The change of one chemical into another by light is called photolysis, and has given rise to an area of science termed photochemistry. Photochemical reactions are the basis for photosynthesis, the mechanism by which plants make food. Whole technologies have been based on photolysis. The blueprint is just one of them and no architect or engineer would be without one only a few years ago. The modern equivalent of sunlight is a high-powered ultraviolet light bulb, but the principles are the same.

Many older readers will recall a popular art form based on the direct photolysis of a colourless material into something visible. Print-out-papers relied on the Sun's energy to produce silver directly in the emulsion. The images were made by contacting a negative onto the paper and leaving the frame in the Sun. Exposure times of minutes to hours were needed depending on the make and type. The resulting image could only be viewed in subdued light (as room lighting eventually fogged it completely), or by having the print fixed in the usual way. Some of these images were exceedingly beautiful, exhibiting a whole range of image colours, very difficult to reproduce by normal photographic methods. There have certainly been serious suggestions that the Turin Shroud was made by a print-out process. We will see later how to use this historic effect in a delightfully simple way to photograph the Sun on a normal day and in eclipse.

The sunlight's effect on silver compounds was well known in the eighteenth century. The Swedish scientist Karl Scheele was one of the first to show that the black colour was due to metallic silver, and by 1784 people were turning their faces black using the idea! However, it was not long before some useful science came along. Johann Ritter in 1801 was able to show that there was another invisible region beyond the violet in the Sun's spectrum, a year after William Herschel had already demonstrated the heating effects of infrared. Ritter found no heating effect, but the more interesting fact that silver chloride was decomposed some thirty times faster than in the visible yellow light. Without this major step in solar study, perhaps the whole science of photography would have been delayed for a very long time.

Images (by then referred to as a photograms) took several hours to form and were not stable. Producing a permanent image had to wait until Nièpce applied an earlier discovery of Senebier that some materials were hardened by sunlight. Bitumen was one such material, sometimes called "Pitch of Judea". A thin coat painted onto a pewter plate hardened where the Sun's image fell on it. The unexposed bitumen remained soluble in a solvent and was washed away to leave an insoluble image on the plate.

Nièpce's first examples of this technique date from 1826. The precise way in which he produced this permanent image remains a mystery, since the Sun's shadow appears not to have moved during the eight-hour exposure period! Viewing the image was difficult. Although pewter had a shiny surface, black images did not show up well against it. Today we call that a poor contrast effect.

Nièpce's first improvement was to subject the images to iodine vapour and then to "fix" them with silver compounds. The next improvement was a movement in the direction of familiar chemistry. He used a silvered copper plate in place of the pewter, and the silver was blacked by treating the exposed portion with iodine. He also used a much improved camera. A partnership with Daguerre proved to be the next major breakthrough.

The Development of Daguerreotypes

In 1831 Daguerre showed that the iodine-treated silver was itself a much more sensitive medium. A few years later in 1835 he discovered the "latent" image effect in which an exposure much too low to make a visible blackening could be "developed" with mercury vapour. The final step into what we would understand as photography took place in 1837, when Daguerre showed that the image could be "fixed" in a common salt solution. Exposure times were now a matter of 20 minutes.

The first attempts at astronomical photography were made at the Paris Observatory in 1839 by its Director, François Jean Arago. The images were of the Moon, not sunlight, and too blurred to be useful because of the low lighting. The first really successful astronomical photograph was made in New York, a year later. An American physicist, John Draper with Samuel Morse (of Code fame) succeeded in photographing the Moon – a small lunar image about 25 mm (an inch) diameter was captured in just 20 minutes.

In 1843 the same pioneering Draper photographed the solar spectrum with the Fraunhöfer lines. Antoine Becquerel in France had independently recorded Fraunhöfer lines on daguerreotypes (as the final images had become known) earlier in June 1842. What is truly remarkable is that not only were both men able to show not only the familiar Fraunhöfer lines in the visible region which any astronomer could see with his own eyes, but also a much more significant fact. The lines could be recorded well into the ultraviolet and infrared regions, totally invisible to the classic astronomer.

The years 1842–43 were truly momentous for scientific achievement and discovery – though this period is rarely given the prominence it deserves. It should be regarded as the real start of true scientific photography. It is also our first example of true solar photography. Draper's achievements are amazing considering the extremely low sensitivity of the medium he used and that the equipment was not specially designed for the work. Even today there are few amateurs who contemplate recording the solar spectrum at any time.

In 1837, John Draper's son, Henry was born. He was to become a great pioneer of photographs, but it was not until 1872 that technology came of age. That period of time was necessary for photographic, telescopic and spectroscopic apparatus to improve to an extent that the first stellar spectrum could be recorded photographically on a regular basis. Henry Draper and his many other cataloguers are justifiably revered as the initiators of such work, but has history credited the wrong generation Draper as the real pioneer?

The first eclipse photograph happened at a surprisingly early date, on 8 July 1842. It was made by the Italian astronomer Giovanni Alessandro Majocchi, a professor in Milan. His daguerreotype took 2 minutes to expose and only recorded the partial phase just before totality, failing to record the corona during totality. Nevertheless, this outstanding achievement ought to be given its due credit as a first.

Within a very short time the technology of daguerreotypes had spread worldwide. They were made much more sensitive with different sensitising vapours such as iodine–bromine or iodine–chlorine. There have been modern attempts to duplicate these experiments, but as you might expect, today's safety restrictions are quite a problem. These sensitivity improvements of 20 to 30 times really did open up the science of photography and it has never looked back.

Today we take short camera exposures for granted but in 1845 there was no such thing as a fast shutter. Portraits took minutes to expose or were "frozen" by huge amounts of flash powder. A fast shutter was needed – otherwise the Sun's image would be terribly overexposed even with the slow daguerreotypes. Hippolyte Fizeau and Léon Foucault, major scientists in their own right, made this major breakthrough which can really be said to be the start of camera exposures times familiar to us. In 1845 their pictures confirmed that the limb brightness of the Sun is indeed only half as bright as the centre. The science of solar studies was given a further boost from this second example of photographic image-making and equipment coming together to make a new discovery.

Our digression into the history of photography has indeed turned out to be of fundamental importance. Solar photography really is pivotal in this fundamental way in making a new discovery and confirming it in a permanent record, rather than the naked-eye observation and the quill pen, ink and paper. Such basic scientific methodology probably came much earlier than most people realise.

The much faster daguerreotypes quickly found applications in what we would regard as the mainstream of astronomy. The astronomers William and George Bond at Harvard were the first to get a really good lunar photograph in 1849, using exposures as short as a minute, with the newly installed 38 cm (15 in) refractor. A year later they got the first stellar images. Although somewhat blurred, they were sufficiently good to show Castor as a double.

Photographing Totality

Solar photography can be said to have come of age a year later in 1851. A total eclipse of the Sun took place in July, passing through Sweden and what was then East Prussia. Two daguerreotypes were taken by

Berkowsky through what even then would have been regarded as modest equipment – a 60 mm ($2\frac{1}{2}$ inch) aperture. The pictures showed the corona.

There is an interesting twist to this historical tale. The coronal intensity was measured and analysed, but not for a hundred years. The pictures had been kept in the Swiss Federal Observatory until 1951. These two pictures thereby have a double tale to tell.

Not only are they the earliest eclipse pictures, but they prove the point that all astronomical observations should be retained and kept in their original format. One simply does not know what technology will turn up in the future to access hidden information locked up unknown to the experimenter. Without the retention of thousands of photographic plates taken through different filters, David Malin would not have been able to make those fabulous colour pictures we see today. These changing celestial objects taken a century ago, long before true colour photography, contain undreamt of data which would not necessarily be recorded for posterity by the modern astronomer with his electronic means. A hundred years hence, will the original electronic machinery still be working to access the original records?

The technology of photography underwent further revolution with the work of Fox Talbot, John Herschel and Frederick Archer to transform it into something we would recognise. This revolution was the ability to produce as many copies as needed. Daguerreotypes were always very fragile originals. Their technique's main technical advantage over the newer competing technology was its superior resolution and contrast, something which is still marvelled at today. Archer's technology led to the collodion process and the now familiar glass plates. A few more improvements later and these plates had become a further ten times more sensitive than daguerreotypes, so that the latter rapidly dropped out of regular use. In 1854, both technologies were used to record an annular eclipse in America.

In spite of the superiority of the collodion process, it had a major drawback in that accurate measurements could not be made on the fragile and flexible surface. The daguerreotype had an interesting revival two decades later in 1874, at another special type of phenomenon, the transit of Venus, when the intention was to make accurate measurements and timings to determine the Earth–Sun distance. Jules Janssen's special revolving or multiple time-lapse image photograph has survived.

Even with Archer's wet plates, which as their name suggests had to be exposed whilst still wet, the technical advantages of their greater speed and multiple copying led the way to the next major breakthrough in eclipse photography in telescope and optical design.

Warren de la Rue, better known as a businessman and head of the company making banknote paper, immediately tried out the new plates. However, he first had to make himself a special telescope to take them. His 33 cm (13 in) reflector had no drive, and he became the first person to experience the frustrations of manual guidance to follow celestial objects. Nevertheless he was able to make some lunar pictures in exposure times from 10 to 30 seconds. Further work had to wait until he had built a much better observatory and a reliable clockwork drive.

Similar problems beset the Americans, and it was not until 1857 that the first really good stellar images were obtained. It was also in America that the optical problems with photography were solved.

Lens Technology

It had been known since the beginning of telescope making that the colours of the spectrum came to a different focus. For visual work this did not matter too much, as long focal ratios eased the problem a great deal, and the eye simply focused on the sharpest image in the colour it saw brightest. Photography changed all that, as much faster optics were needed for the short exposure times. The collodion plates only responded to blue light, and of course the violet and ultraviolet which the eye could not see. Accurate focusing became a really knotty problem.

Achromatic lenses in the 1850s were designed for the eye's maximum sensitivity in the green to orange. Lewis Rutherford in New York discovered that his telescope for maximum photographic sensitivity focused 2 cm (up to 1 in) away from the visual image. No matter what he did, it was impossible to get a really sharp image, as the other colours contributed something. Much work and labour were needed by Henry Fitz, an established optician, before a photographic achromat appeared in 1864.

The task was incredibly difficult. Just imagine the problems: Fitz was trying to make the lens focus but was unable to see the image. Each repolishing needed the preparation of a wet plate and exposure on a bright star and the result had to be intensely scrutinised until the sharpest image turned up. The result was a 28 cm (11 in) objective, and with this Rutherford was able to record stars fainter than the naked eye could see in the sky.

The equipment was crucial in the history of astronomical photography, and in general photography because of the attention to colour corrections. More importantly, it started the science of stellar measurements and proved for the first time that accurate positions can be achieved. Lunar pictures became accurate enough to allow significant enlargements with exposure times as short as a quarter second. After these major achievements, Rutherford dropped out of astronomical research to concentrate on other endeavours.

The 1860s

Whilst this vital piece of research was going on to improve lenses, de la Rue had continued his research with mirror optics. Mirrors do not present chromatic problems, since all wavelengths come to the same focus. Much larger mirrors are needed because of the low reflectivity of metals, speculum in his case. His new observatory was based at Kew.

John Herschel was also interested in another long-term project, and suggested regular solar photography. De la Rue constructed his new telescope, a "photoheliograph", probably the first of a new generation of telescopes dedicated to specific projects. With ample light around from the Sun, this time he used a lens with a modest aperture of 9 cm $(3\frac{1}{2}$ in$)$ which was much easier to make than the larger Rutherford design.

One of our famous historical eclipses occurred before de la Rue could start on his patrol work. He took the equipment to Spain in July 1860 to collaborate with Father Angelo Secchi from the Collegio Romano to settle another astronomical dispute of the time. This was the theory that solar prominences were terrestrial atmospheric phenomena.

Secchi stationed himself 400 km (some 250 miles) away with a more conventional 15 cm (6 in) visual refractor. Both workers obtained several pictures during totality which proved conclusively that the prominences were part of, i.e. attached to, the Sun and definitely not a lunar phenomenon.

This very neat and crucial experiment can be said to be the start of all pioneering solar studies. When one considers the difficulties of carrying out the experiment with wet plates, it speaks volumes for the value of meticulous planning and practice, and luck in having clear weather at sites 400 km apart.

The de la Rue instrument returned to England and formed the centrepiece of the Kew Observatory. One of its first tasks was to measure the Earth's magnetic field as it had become clear by that date that sunspots affected it in some way. The monitoring work continued for a full solar cycle before better new equipment appeared on the scene and the work passed to professional hands at Greenwich. The linked history of photography and solar studies does not end here, however.

Solar Spectroscopy

We have already seen that Fraunhöfer lines were amongst the first phenomena to be recorded by Draper in 1841 and well into the ultraviolet and infrared. The history of spectroscopy is a separate study on its own; suffice to say that a large number of Fraunhöfer lines had been identified by the mid-1860s. Workers with famous names such as Bunsen, Kirchhoff and Ångström feature. For many years there was considerable doubt what some of the lines indicated. One yellow line remained a puzzle, and was assumed to be sodium. By 1868 no less than four eclipse expeditions were geared up to measure the spectrum of the newly photographed prominences. Nearly all the lines were easily identified with hydrogen, but there remained the unidentified yellow line which could not be resolved from sodium with the crude instruments then available. However, it was obvious to one observer, Jules Janssen, that whatever the material was it produced an emission, not an absorption, line, and it should therefore be visible during any day, not just during an eclipse.

By one of those curious twists of fate that happen from time to time, the same concept had occurred to two different scientists who were not even present at the 1868 eclipse, Norman Lockyer and William Huggins. They published on the same day as Janssen.

With the aid of more accurate apparatus, the wavelength of the "new" yellow line was measured, and proved to be different from the sodium pair. As nothing on Earth fitted the line, and it was so strong, it had to be a new element. It was given the now familiar name of helium in 1869 by Edward Frankland, a chemistry professor at Imperial College.

This application of spectroscopy at a time of eclipse can be said to be the birth of the new science of astrophysics. Without it, the composition and physical state of the stars and the universe about us would never have been possible. Both Lockyer and Janssen went on to become pioneers of a new style of laboratory devoted to the new sciences of astrophysics rather than conventional observatories.

The historical eclipses in 1851, 1860 and 1868 really do have their place as epic milestones in the history of science outside astronomy. Later eclipses built on the pioneering work and expanded into the new sciences of astrophysics when it became possible to measure the composition of distant galaxies, their local conditions and motions with amazing accuracy.

Eclipse studies did not all have such a happy ending. For the best part of the next century some spectral lines refused to be identified, even with the "new" element helium, and some other elements were suggested with the fanciful names of nebulium (found mainly in nebulae outside the solar system) and coronium in the solar corona. When radioactivity was discovered it was clear that neither was a true element.

The mystery was only solved in 1941 by Bengt Edlén in Sweden, when the true high temperature of the corona was confirmed to be about a million kelvin. At the exceptionally low pressures there, iron atoms stripped of no less than nine electrons are sufficiently stable to radiate at the characteristic pink colour of the inner corona so familiar at eclipses. With a wavelength of 637.4 nm, this is sufficiently close to the H-alpha line to be confused by the naked eye. However, it will not show through specialised H-alpha filters.

Another coronal line is due to iron losing 13 electrons. It radiates in the green. Both the red and green lines are very intense, and are one of the reasons why accurate photography is difficult because of the severe halation and focusing problems through lenses.

Before leaving the inter-related topics of astronomy and photography, let us note that the Sun still has a part to play.

The spectrum wavelength standards became increasingly important, and the primary reference standards were not terrestrial but solar. Accurate records could be made with the newer plates and distributed. The work of Henry Rowland at Johns Hopkins University had the necessary precision and was adopted for a long time. His measurement of thousands of Fraunhöfer lines is still a model of persistence, particularly as he repeated the task. It is hardly surprising that he became the university's first physics professor.

Janssen returned to the scene again in 1885 when he obtained a striking photograph of the Sun's granulation, only briefly seen in moments of exceptional seeing. Even as late as the 1950s, reference books still referred to Janssen's picture as the best of solar granulation in existence. Many workers today with all the technological improvements since 1950 still find it a challenge to come close to Janssen's quality. The term granulation was adopted as more scientific than the "rice grains" or "willow leaves" phrases used up to then.

The final chapter in our present story is the major improvement in the spectral sensitivity of the photographic plates themselves. The need to record the infrared spectrum became ever more necessary. Captain W.W. Abney, more famous in spectrometry, worked for years before perfecting a plate sensitive to this invisible radiation. He was able to achieve this long after the collodion plates had changed to gelatine, familiar today. His red-sensitive plates opened up the whole field of panchromatic photography, ultimately making full-colour photography a reality.

Chapter 7

Partial Eclipses

At a partial eclipse, the Moon may cover anything from a tiny fraction of the Sun's disk up to more than 99%! Unfortunately, even a very large partial eclipse is not sufficient to reveal the Sun's surroundings; for this, absolute totality is needed. And always remember that the partially eclipsed Sun is just as dangerous as the Sun at a time of non-eclipse.

As the tables in Chapter 13 show, a partial phase is seen over a wide zone to either side of the track of totality, but there are also partial eclipses which are not total anywhere on the Earth's surface.

The first sign of an approaching eclipse is a tiny "nick" in the Sun's brilliant surface; this is termed First Contact. Gradually the Moon's disk encroaches onto the photosphere, and if the eclipse is a large one there may be an appreciable reduction in sunlight and heat, though this is not usually very marked until at least half the Sun is covered (see Fig. 7.1). After maximum, everything happens in reverse order, until the Moon passes off the solar disk at Last Contact.

Naked-eye observation can be carried out, using the precautions detailed in Chapter 2, but much the best method is to use a telescope (preferably a refractor) to project the Sun's image on to a suitable screen or card behind the eyepiece. It is always worth making a series of sketches as the eclipse progresses, and of course to take photographs at regular intervals.

If sunspots are on view, compare their darkness with that of the Moon's disk. The result may seem a little unexpected; the Moon is much the darker. As we noted in Chapter 2, a sunspot is not really black; it appears so

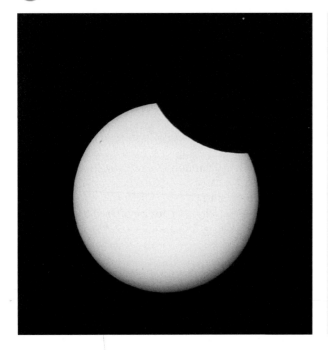

Figure 7.1. A typical view of any partial eclipse of the Sun. (Patrick Moore)

only because it is some 2000°C cooler than the adjacent photosphere. If a spot could be seen shining on its own the brilliancy would be greater than that of an arc lamp – whereas the disk of the Moon really is jet-black.

A partial eclipse is always worth watching visually; look for instance to see if you can make out any irregularities on the Moon's limb due to lunar mountains or valleys – for this, make the projected image as large as is practicable. But all in all, there is no doubt that to the "casual" solar observer the main interest of a partial eclipse is photographic.

Partial eclipse photography is no different from solar photography on any normal day. Almost. The excitement at an eclipse is something special. Being prepared with well-tried equipment underscores the points we keep repeating. Preparation and practice. Why spend enormous amounts of time, effort and money to record the rare events, then let the equipment lie fallow for years, or discard it altogether? It makes sense to make regular solar observations and get multiple value out of the expenditure.

Partial eclipses cover a much larger part of the Earth than totalities and are more likely to be seen from home. All the more reason to start from here in a familiar place and with everything to hand.

Photography or Viewing?

Safety must be paramount. Read carefully the warnings and advice in Chapter 2 before making a choice. Obviously, projection has to be the simplest option for naked-eye work – and the safest. The more ambitious can then use any photographic method to capture the projected images. But don't be too ambitious. Certainly never try to be clever on your first go. Even experienced photographers, the authors included, make ghastly mistakes. The best advice offered to newcomers is to regard photography as a bonus, whether with a "steam" camera or the latest video technology. Be psychologically prepared to view the eclipse with nothing more than a filter and the naked eye.

The supreme advantage of naked-eye viewing is the complete lack of clutter, which is almost as important as no last minute changes of plan.

Any experienced eclipse chaser will tell you that the most vital thing to do is to make plenty of time to just stand and stare. Study the lighting effects in the landscape around you, and how plants and animals behave. You will keep the mental picture for ever, and no photograph will ever replace that for vividness.

Photographers must always come mentally prepared to ditch all their cherished plans (and equipment) and rely on simple naked-eye viewing. Someone, somewhere is bound to get a good picture which you can keep as a souvenir. Snapshots of the site and people present can also have some sentimental value later, particularly if someone does something bizarre.

Some Basics

Solar photographers rapidly bump into some unpleasant basic problems. Not least of these is "running out of focal length".

The longer the focal length, the better the maximum detail. Terrestrial optical equipment is designed for the totally different purpose of enlarging what the eye sees by a sensible amount. Longer focal lengths cause unsteady images because of atmospheric turbulence, or your inability to hold the camera steady. Somewhere there has to be a trade-off to get a decent image size and steadiness.

The solar disk size is often quoted as 0.0093 of the focal length. We can save much mental anguish by using the simpler formula of 1% of the focal length. But what does that mean in everyday language? Focal lengths over one metre (one yard) are needed to get a "useful" image size. A crude definition of "useful" has to be an image size which appears similar at a normal reading distance to the quick glimpse of the Sun in the sky (something you must never do!). This places it firmly in the telescope range, beyond binoculars and way outside the range of common camera telephoto lenses.

Which brings us straight back to projection. For partial eclipses and sunspot checks this is by far the best answer anyway.

Solar Projection

Whatever you do, consult an expert in a local society or reputable astronomical equipment supplier before buying something "recommended" for solar projection. That advice also applies before adapting something used for more conventional night-time astronomy.

Direct solar projection out of the "business end" is the configuration most commonly seen and sold. Adopt it wherever possible, because the solar image is orientated properly and everything can be immediately identified with the real thing. It is only in the special case of solar photography that straight-through projection becomes difficult.

The first awkward point is that cameras get in the way of the telescope, and vice versa, and cannot be placed in the best position, which is square-on to the viewing screen. Then your head will get in the way, unless you go well back, in which case the image gets smaller, and so on with increasing complications. All these niggles are soluble, but they do require thought and special attachments.

The alternative of photographing through the back of a screen is little better, because of the degraded image. Also, there is always the risk of looking straight into the Sun.

Sometimes the best solution is a telescope with a right-angled eyepiece attachment to project the solar image sideways. The classic Newtonian is designed that way already and is perfect for the job.

The most obvious immediate benefit is that you can project as far away as you like without the full blast of sunlight leaking round the telescope. The image can be projected onto a white wall, for instance, or through a window into a (relatively) dark room. A mini coelostat, if you like. The further the projection, the easier it is to get a camera square-on, and as many cameras as needed. These can be placed to get the best magnification or clarity.

What problems can you anticipate? Well, quite a few really.

The longer the projection distance, the more precise the mount and drive have to be. Hand holding works fine for small solar images. Nevertheless, the reflection(s) introduced by the angled eyepiece test your IQ and manual dexterity, particularly with the hand holding option. Then there is the knotty problem of sorting out the correct orientation of the photographed images afterwards. The best way round this is to take accurate notes or persuade everyone that you prefer the crescents that way up!

Snags notwithstanding, the sideways projection method is the easiest to organise for cameras and videos which will be used exactly as supplied, i.e. without modification. Film or tape is quite immaterial, simply use whatever is in the camera or video already.

Relays of photographers can be fed in and kept away from the more serious experiments elsewhere, always an important consideration. There is also less risk of jogging the telescope at a vital moment.

There is absolutely nothing wrong with the straight-through option, which will be chosen by most readers because of its simplicity and easy disk orientation. Always use a projection method for public demonstrations, so that everyone in the crowd can be assured of "their" own picture (see Fig. 7.2).

Pinhole Projection

Our basic piece of kit is entirely natural. Nature has provided us with a huge range of varieties – trees and bushes! On a normal day, the Sun's image cast through small gaps in the leaves is impossible to distinguish from the more general lighting flooding through bigger chinks. As the eclipse progresses, some shadows will become more clearly circular and crescent shaped.

Figure 7.2. Dan Turton projecting the Sun's image against a wall to allow others to observe and safely photograph the eclipse. This photo also demonstrates that better contrast could be achieved by using a black screen instead of white. Santorini, 1976. *(Michael Maunder)*

These make a marvellous sight, and become more and more distinct if the eclipse approaches totality.

Now is the time to photograph the eclipse with any technique you like. Lighting levels will be close to normal, and the simplest of cameras should be capable of recording the scene with no difficulty at all.

Here we must make a strong plea for common sense. Never use a camera with flash at any time during an eclipse. Not only will it wash out the image but it will infuriate other observers trying to make serious observations. If all you have is a simple camera which cannot deactivate the flash, do everyone a favour and leave it behind.

These natural pinhole images under trees and bushes are tailor-made for camcorders. Electronic recording works very well at partial eclipses and owners will not need to consider more conventional options except as a back-up or for higher resolution.

Camcorder electronics will cope throughout with the wide range of exposure conditions. Use a wide-angle lens and include a lot of background scenery in a close-up zoom sequence. The gentle movement and dappling is a sight to behold, and invariably shows up well on replay through any domestic TV.

What if no natural pinholes are around? Go back to medieval times and find a nice, large dark room with a blackout window facing the Sun. A chink in the black-out will cast an image of worthwhile size on the far wall which can be photographed with the simplest of "fun" cameras to the latest electronic camcorder. Any dark-ened space such as the inside of a vehicle can be pressed into service on location.

A temporary "darkened" room can be made with a little ingenuity and a large cardboard box. Photography is possible through any hole cut in the side or top.

Another simple idea is also the safest to make on the spot. Take a large piece of cardboard, the larger the better, to form its own light baffle, and pierce a small hole in the middle. Hold this up to the Sun and an image will form on a white paper screen which can be photo-graphed with any equipment you like (see Fig. 7.3).

Remember, all you are doing is recording the scene viewed with the unaided naked eye, something a camera is designed for.

Although this sounds terribly crude, it is a remark-ably effective way of starting solar photography. The larger the degree of eclipse the more effective it becomes as the ambient light levels drop, but the solar image retains its brightness.

In full sunlight a distinct solar disk is distinguishable up to several feet between screen and pinhole. The wider you can make the light baffling, the wider this separation can become. With dark cloths used by Victorian photographers looking at their viewing screens, some surprising focal lengths can be achieved. Bizarre dark cloths range from ordinary clothing to black dustbin liners.

Pinhole technology is great fun, and by far the best teaching aid because of its built-in safety factor of only viewing a projected image. Although the final images can never be as good as through optics proper, eclipse

Figure 7.3. Using a simple piece of cardboard as a light baffle.

progress can be monitored by a large number of people. In a permanent set-up, changes in the solar disk size during the year are quite noticeable. The equipment costs nothing but a little time and thought and, more importantly, the system works without any special solar filter. Nor do you need one for the next set of simple ideas.

Pinhole Cameras

Our next ambition is to make a proper camera. All the conventional photographic kits examined by the authors have much too short a focal length to be of any real value in solar photography, and so it is back to the drawing board to make one.

Warning! We cannot recommend the simplest idea of all. DO NOT use a long black box or tube with a sheet of greaseproof paper flat at one end and the pinhole at the other, and point the pinhole end of in the general direction of the Sun. Even with a large

Figure 7.4. Peter Cattermole making use of everyday objects... using a cowpat as a pinhole projector. Arizona, 1994. (Michael Maunder)

baffle around the tube, the risk of looking straight into the Sun is simply too great.

In our discussion of the history of photography in Chapter 6, printing-out papers were mentioned. Any commercial photographic paper (particularly dirt-cheap-out-of-date-job-lots) will do. Our camera must be light-proof; the photographic paper is loaded in the dark and takes the place of the viewing screen.

Exposure times are too long for any image to be registered in the second or so before the Earth's rotation blurs it. In the absence of a polar drive, keep the exposure times really short and develop the paper in the normal way.

The next logical step is to attach a film camera to the business end. The camera does not need to be a good one; indeed it should not be, as all we want is something to hold the film in place (and flat). Car boot sales are an infinite resource of vintage cameras with faulty lenses and shutters. Just remove the faulty parts and glue/fix the film holder to the business end. Plastic-bodied "Brownies" are just made for the job! If the shutter still works, keep it, remove the optics and use that as the pinhole, otherwise exposure times can be controlled by the "Mexican hat trick".

Conventional camera backs are a huge advantage because many more exposures are possible before reloading, and a wide choice of film can be used. Since this is the economy route, the very cheapest out-of-date film is recommended.

A once very popular film size is 126. Refill cassettes can still be obtained and pressed into service. Hours of endless amusement will then be spent trying to get the Sun accurately aligned so that its image falls only onto the minuscule film size. It can be done. Experiment with a sighting device made from a shadow caster. Obviously, the larger the film format of the original camera, the simpler the task becomes.

After many years of neglect, the "fun" camera has come back into fashion. Whilst these should be re-cycled, the majority are abandoned and become an endless resource for pinhole photography. You will need two for each pinhole camera.

Simply split the plastic bodies (most come apart, if not easily, with some "persuasion") and use one as the lens/shutter and the other as the film end. Remove the lens and any spacers from the shutter and replace with a small piece of metal foil with a hole in it. A few minutes' study will work out how to re-cock the shutter with a matchstick without operating the film wind-on mechanism. A few minutes more and the new shutter mechanism can be glued at the front of the camera.

The film end is a little more complex, and some radical surgery is necessary to remove the lens/shutter unit without upsetting the film wind-on mechanism. The majority of the "fun" cameras take ordinary 35 mm film cassettes, and it is not too difficult to work out how to reload each particular make yourself.

The great advantage of a pinhole as our first entry into solar photography is that the image comes into focus no matter where the screen is. In the specific case of the Sun, halation enlarges the image to separations much under a hand-span. The image is of uniform size, and then begins to expand noticeably the further away you go. The image is always a little blurry around the edges because of diffraction effects. As a teaching aid and to get one started that does not matter, but the next quantum leap comes when we move up to lens optics.

The "fun" camera is our starting point once again.

Simple-lens Photography

Any of the pinhole cameras we have made can be improved out of all recognition by the addition of a simple lens. This biggest disadvantage is that the image is formed only at the true focus of the lens, and that the film has to be placed at that precise distance. It requires quite a lot of trial and error to get that distance right, but once set you don't need a bellows or similar complication to focus on terrestrial objects.

We have already discussed the focusing problems of simple lenses (Chapter 6). No single simple lens is capable of bringing all colours to a focus on a flat film, and even the eye cannot adjust enough. For solar disk and partial eclipses this does not matter too much, since we only need an image of some sort in any old colour.

We will take a leaf out of the book of those "nasty" cheap telescopes on the market which use a simple lens and a stop, to get any sort of performance at all. The smaller the stop, the better the colours focus together, but the terrestrial image gets too faint to be of any use. However, we are in luck: we have light in abundance.

Pick up one of these telescopes cheaply and use its mount and camera attachments with minimal modification, then set about improving the lens performance. Araldite the shutter and iris portion of a "fun" camera to a cap fixed rigidly over the lens, and as close to the lens as possible. Remove the lens hood if necessary. The average iris size of a "fun" camera is a mite too large for a true pinhole, but ideal as a stop. With such a massive stopping down, the lens should focus quite well. The iris can be drilled out larger if more light is needed since the stop will be considerably better than anything supplied with the telescope.

Remember, this is only a dodge suitable for good light situations at partial eclipses, and is no answer for annulars and totals, when we do need proper optics.

What if we have no cheap telescope or body to work with? There are some excellent photographic quality long-focus lenses in every camera shop. These are called "close-up" adaptors, intended for macro photography. The close-up 1 lens is +1 dioptre, which, by definition, has a focal length of 1 metre. They are never quite that, but near enough to start making tubes. The close-up 2 is one half the focal length at 500 mm and so on. Choose the cheaper, smaller diameter ones of 49 mm or less and build a camera around that in the same way.

These simple cameras cost next to nothing, and once focused are quite adequate as a teaching aid, for routine solar monitoring, and partial eclipses. They also become a good stand-by for the more serious work in the next chapters. Focusing them is only a question of perseverance.

Testing the Set-up

Since we cannot safely look at a solar image on a focusing screen in case something goes wrong, how do we do it? The answer is to use the Moon near full. Since we are only interested in eclipses, the image size has to be the same, and distance is still infinity. No messing about with fuzzy images of telegraph poles or the like on the horizon.

Instead of using the "fun" camera immediately, drill a small hole of about the same size as the iris in a metal sheet and place that over the lens. That will solve the problem of the very short exposure times from the average shutter. If B or T was a feature of the old camera, use that straightaway.

Lining-up and centring the image on a focusing screen of ground glass can now be worked out in complete safety, and with the added advantage of no distracting bright sunlight. Any chromatic, spherical and other aberrations will show up immediately, and unsuitable lenses can be rejected immediately. The minor difference in diurnal rates of the Moon and Sun can be also be ignored if you need to check out drives and mount stability at the same time.

Once the lens and centring problems have been sorted out with the naked eye and a focusing screen, the camera proper can be constructed knowing the precise focal distance, and then checked out with test exposures. Moonlight needs too long an exposure with our huge stopping down, so jump straight in with solar exposures now that everything has been lined up. All so much easier than risking eye contact with the Sun.

Our final venture combines projection directly with a normal camera.

Solar Projection

Projecting the Sun with a telescope or binoculars described elsewhere uses the instrument exactly as it comes out of the shop, the only difference being the image formed on a screen some distance away rather than in the eye. Afocal projection takes the image outside the eyepiece and uses the camera to bring it to a focus on the film. The process can get a little complicated, and we will discuss only two ways of doing it.

First, the warnings given in Chapter 3 are very important – use only top quality solar filters over the lenses.

Direct Projection

Any SLR camera without a lens is fine for our purpose; all we need is some device for fixing the eyepiece end of the telescope to the camera in perfect alignment. The telescope is then brought into focus in the camera viewfinder, and test exposures made to get the optimum exposure. If the image fills the screen the camera metering can be used as a rough guide.

The main advantage of this technique is that an infinitely variable image size is possible, and huge enlargements of segments of the solar rim for Baily's Beads, or close-ups of sunspots, are no trouble at all. It is also one of the finest techniques known for resolving the problem of camera shake completely. Provided there is good light exclusion round the "joint" there need be no physical contact between the camera and telescope. Amplification of the shutter vibration is not carried through the optical train, and the camera records a static image, which it is with no physical linkage.

The principles behind this simple concept are much underrated in the literature, and deserve a better hearing. The advent of aspheric optics vastly improves the image which will focus quite well then at all wavelengths. The lack of accurate colour focus has been the major problem up till now.

Afocal Projection

The major difference is the use of the camera lens, usually the standard one. The system is so much easier to focus and is done for you on the camera lens (see Fig. 7.5). The principle is very simple indeed. The telescope is focused by eye to infinity, which means that light emerges parallel. Night-time objects such as the Moon or stars will do for this job, and the focus position marked for future reference.

The camera lens has a mark for infinity focus, which is normally at one end stop. At this point it focuses on light coming in parallel. The upshot is that the distance between the telescope eyepiece and front of the camera lens does not need to be intimate, allowing a surprisingly wide separation. As with the direct projection option, direct contact is not obligatory, and camera shake disappears completely. If the camera has a leaf shutter instead of an SLR mirror, camera shake is effectively zero to begin with and can be ignored.

The main advantage of this set-up is the complete control over what you are doing, and ability to split it up again after use and return the pieces to their original function. This has to make sense on expedition if the telescope is, in fact, a pair of binoculars.

The "power" or magnification of the telescope becomes the increase in focal length of the camera lens. The effective focal ratio is this new focal length divided by the telescope objective size. If we use 10 by 40 binoculars as an example with a standard 50 mm lens on a 35 mm camera, the new focal length is 10×50 mm = 500 mm, quite an improvement and very usable.

The effective focal ratio of the camera lens is irrelevant, and is now 500/40 or 12.5. The next highest conventional f-ratio is used instead, and this is $f16$.

Figure 7.5.
A telescope with camera attached at right angles for afocal projection. Arizona, 1994.

We use the next highest f-ratio because there is some inevitable light loss through the extra glass in the system. This light loss is a small price to pay for convenience, and as we are only using the very central portion of the imaging system, image quality is surprisingly good. One author (MM) has used this system for several eclipses and obtained sharper images than with conventional optics, because of the lack of camera shake. A leaf shutter and highly compact construction are the key and this equipment has been good enough to capture front page cover of total eclipses in magazines! There are many snags to the system, but the practical advantages far outweigh them for partial and annular eclipses, and totals with caution.

The biggest drawback is the marked colour shift due to the amount of glass in the many lenses. This can be corrected very easily with modern electronic technology.

All the other snags are also optical. Modern high speed (low f-ratio) camera lenses are a distinct disadvantage as the image vignettes severely with them and a circular field of view results. The older and smaller diameter the lenses are, the better, preferably similar to the exit pupil of the telescope. Only the very central portion of the image is anything like in focus. A short distance off the optical axis and the image quality breaks down completely and becomes unusable. The technique is quite unsuitable for coronal studies, for instance.

Because of this dependence on accurate centring, very precise alignment is vital. One answer is to use binoculars on a twin-lens reflex camera so that the viewfinder image is the same size.

Focus accuracy is also highly dependent on the telescope focus and eyesight is not always perfect. Anyone attempting to make this simple equipment is strongly advised to carry out plenty of test exposures. The Moon is an ideal test subject and with an effective f-ratio of 16, by definition derived from daylight exposures, 100 ISO film needs 1/100 s exposure as a starting point at full moon.

Practise and practise these simple ideas until they are second nature. Although the apparatus is very crude, it is perfect training for the real photography to come.

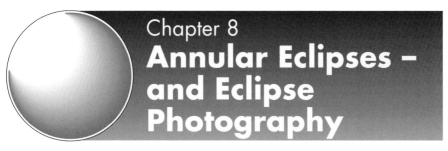

Chapter 8
Annular Eclipses – and Eclipse Photography

Annular Eclipses

Annular eclipses are more frequent than total eclipses, simply because the average length of the Moon's shadow is not quite long enough to touch the surface of the Earth, but from any particular locality even annulars are not common. Over England, the last was in 1858; the next will not be until 2093.

There are some eclipses which are termed "annular–total", with the shadow touching the Earth in the middle of the track and missing it to either side. Obviously, this means that even in the most favourable location the duration of totality will be very brief.

At an annular eclipse where the Moon is very nearly large enough in the sky to cover the bright surface of the Sun; only a very thin ring is left showing round the dark lunar disk, and it might be thought that at least some of the phenomena of totality could be seen. Yet in general this is not so. Baily's Beads are visible, and are well worth photographing, but there is no corona, and no prominences flash into view. Neither have we any reports of shadow bands being seen during annularity, though admittedly there seems no obvious reason why not.

The darkening of the sky is evident enough, and so is the drop in temperature, but neither effect is as marked as might be expected. In 1976 there was an annular eclipse visible from the Mediterranean, and both authors observed it from the island of Santorini – the island which, according to tradition, exploded so violently around BC 1500 that the tidal waves wiped out the advanced civilisation on Crete. Even in mid-eclipse, many of the local population took no notice, because they had not realised that anything unusual was happening.

There is one vitally important point – which, without apology, we stress yet again. During annularity, the Sun is just as dangerous as it is when uneclipsed, and all the usual precautions have to be taken. It is only when the photosphere is completely hidden, and the corona has come into view, that it is safe to observe the Sun directly with any optical equipment.

This being so, what observations should be carried out? Certainly an annular eclipse is a fascinating spectacle, but whether any useful scientific work can be done is quite another matter.

Photography

The challenges which an annular eclipse present to the photographer will be covered in some depth in this chapter, since the technology needed is the same for all eclipses up to totality.

Unlike a total eclipse, where the sky conditions have to be perfect, partials and annulars have less scientific importance and the social atmosphere is much more relaxed. For instance, there is no need to worry about removing the filter at a critical moment and the exposures do not need so much bracketing.

Much of the time will be spent taking the occasional picture of the progression, and any of the projection methods can be used.

The major difference at an annular is the possibility of much more dramatic lighting effects in the local scenery as the eclipse deepens. To many, this phenomenon is the most important aspect of any eclipse. The effects are very difficult to capture if the "serious" work is being done with expensive cameras in the main experiments. Whatever you do, forget any idea of swapping the expensive stuff about, because something is bound to go wrong. Think laterally.

Fun Cameras

Fill your pockets with as many "fun" cameras as you can. They have been designed for general scenic pictures, unexpected events and people, and cost little more than a reel of film and so work out the cheapest of all to run. They weigh little and, if bought in bulk, come ready packed in neat boxes for expeditions. The modern optics give some quite outstanding results, and film latitude and printing technology cope with an unbelievably wide exposure range.

Fun cameras come in a wide range of types, including panoramics which are often better for scenic shots. Panoramic lenses on a conventional camera cost a fortune, and often do not give better pictures. Fun cameras capture scenes quite impossible with more complex alternatives. They invariably work perfectly, as they are intended to be used by the fumble-fisted. For the peace of mind they create, they are truly worth their weight in gold. If the eclipse is clouded out, "happysnaps" of the scene are better than nothing.

It is possible to reload the disposable cameras and use whatever film is preferred for regular cameras. Many High Street processing shops are only too happy to get rid of them to experimenters. The next step up from a fun camera is the cheapest of the "point and shoot" variety which can be reloaded. These can be obtained in a bewildering range. As can the more expensive "rangefinders". These used to be regarded as toys, but modern ones often offer more features than a conventional camera.

The only question remaining is how do we use all these cheap cameras?

Simple, just follow the instructions except that our whole point is to work at low light levels. The film latitude and printing take care of it all. The more complex camera with a single or fixed shutter speed is used in exactly the same way. Because the shutter speed is exactly the same for each shot, the negatives will record the lighting faithfully, useful data in itself. Slide film is by far the best option here, as the processing is rigidly standardised.

It is a complete waste of money and effort to carry tripods for these simple cameras. Use any firm surface to steady the camera and the picture sharpness will improve out of all recognition. A little ingenuity with double-sided sticky tape or Velcro will work wonders with bits of stake hammered into the ground.

The more complex the camera, the more relevant it is to use a tripod for each one. If the camera is auto-exposure only, there is no alternative. There are plenty of very cheap and portable tripods sold today.

Camcorders and video have not been mentioned up to now because their superiority in recording all the phenomena has already been covered. They are the best option for capturing the pinhole effects under trees, and zooming out for atmospheric impact. At annularity, the electronics will still cope even when the event is almost total.

Some Basics

We emphasised the importance of not using flash cameras during a partial eclipse. Do not bring them on to site. Dark adaptation is even more critical during an annular.

Cameras will be sitting in the Sun for a very long time. Most cameras are black and absorb heat very well, which will "cook" film extremely efficiently. Don't rely on camcorder electronics being totally immune either. Make sure that all cameras have adequate shielding from direct sunlight as much as possible.

Some annulars display a wide ring which is little different from the brightness of the main disk. A long eclipse of this kind can be looked upon as just a special case of a partial eclipse, and recorded in the same way.

It is only when the annular is close to a total, or the Baily's Beads are the main interest, that things become serious and nothing but the best equipment will do. Projection with good quality telescopes is still the safest option, but the image becomes increasingly more difficult to photograph near to totality. The simpler cameras with autoexposure only are particularly bothersome, needing good tripods.

The solar limb is at best not as bright as the centre, and markedly less bright at the extreme edge. Large projected images might not be bright enough to record satisfactorily with the simpler cameras, although video should be able to cope. Naked-eye work becomes progressively more difficult, even with good dark adaptation. To record the solar limb properly, and to pick up subtle detail in Baily's Beads there is simply no option but to use the largest magnification possible. Because of the intrinsic faintness, use only the best optical and recording equipment.

Cost is usually the decider for the more complex equipment. However, simplicity is much more important. A totally new camera can be a recipe for disaster unless sufficient time and trouble are taken to get used to the controls, which must be second nature. This is true whatever type of recording medium you opt for.

Some auto-function cameras may suit you for everyday work, but they must be checked out first. This is particularly true with some videos, because they rely on infrared detection for some functions, and the whole point of solar filtering is to ensure that no radiation gets through.

Because large primary images are needed, SLR cameras remain the most popular choice for annular eclipses, but the situation is changing rapidly. Some of the cheaper camcorders spring a number of surprises each time a new one appears, and it pays to keep a weather eye open.

Whichever recording medium is chosen the same basics apply. Check and double check that adaptors and fittings are made and do work when using lenses not designed by the camera maker. Check over battery consumption and buy only fresh batteries with plenty of spares, before you leave home. The same applies to film and tapes.

To recap, the history of eclipse trips is full of tales of photographers failing to get pictures because they have not had a camera long enough to get used to its funny ways. Stick with an old one if in doubt. Better still, take your old camera along as well. All professional photographers carry a spare (or two!).

Film

Video workers have little option with the recording medium, it boils down to buying the best quality that you can. Digital recording is increasingly common. "Drop-outs", "snowstorms" or poor bandwidth are unwelcome on a black background. Images should not need to be cleaned up with computer technology.

Film workers should adopt the same concept. There are no "bad" makes of film these days. Buying the best quality in our context means the slowest speed film you can get away with for the expected shutter speed, which in turn is determined by the solar filter. The exposure range is nowhere near as wide as for a total eclipse because there is no outer corona to worry about, although some marginal annular events will show a trace of chromosphere or even inner corona.

A film speed of 100 ISO or lower ensures the best resolution and contrast. As the image is nearly monochrome, there is considerable merit in using a black and white negative film with Mylar or aluminised filters. The images can be printed later through a filter to make any colour you prefer. The main snag with this process is that the subtle gradation of colour towards the red/orange can be lost, but that is a small price to pay to get away from a hideous blue.

In the special case of recording a timed sequence through Inconel and other neutral filters, the immediate impact of the 50 ISO slide films is dramatic, particularly in the larger formats of 120 and over (see Fig. 8.1).

Glass Optics

Sheer image size is the name of the game for Baily's Beads, and is probably the only criterion. For the Diamond Ring effect, quality has to come into it too, and with a vengeance. Too many pictures are ruined by halation, "ghosting" and secondary imaging to be complacent on this topic. The sheer beauty of the Diamond Ring effect demands nothing but the best optics.

Proper lens blooming is part of the answer, but even that can fail this severe test. Mirrors or filters in the optical chain also fail. It becomes a "Catch 22" situation in planning ahead, as the only safe way forward is to use equipment working properly on a previous occasion. The image-destroying secondary images need the savage intensity to appear, but the following test can be helpful in ditching the rubbish early. Any old film will do, or a reel end. Proceed as follows:

Photograph an intense light at night, such as a street light on the horizon, without the solar filter. Make the exposure at least ten times longer than metering or common sense indicates. At least two exposures are needed. Make the first with the light source dead centre, the other with the light somewhere close to where the Ring is expected on the film frame. Get the pictures developed, and study the result.

Any optics showing ghosting now is going to give a disappointing result with the real thing. Sometimes the ghosting is acceptable or nearly absent with dead-central imaging, but is it worth the risk of rescheduling the exposures?

Do not add tele-extenders until the particular combination has been tested fully. You will get a severe light loss and image degradation, usually manifested as much poorer contrast. By far the more serious effect is the worsening of camera shake.

Avoid zoom lenses, even with modern optics, and buy the best "prime" lens you can afford. Quite often a simple glass lens out-performs a zoom, with better contrast and fewer secondary images. This ghosting problem or multiple imaging is an absolute killer with zoom optics, most common on camcorders and video.

The video detector system also throws up a few problems of its own. Quite often a pixel, or row of pixels, will have a different sensitivity, or overload threshold. Watch out for whole rows going down and displays of "crosses" and "bars" appearing on top of the Diamond Ring, destroying the whole effect. Testing with street lights will rarely show up the fault, because the intensity is too low and the safety cut-outs can work as intended.

Mirror Optics

Mirror telelenses and telescopes are usually much less prone to ghosting and are a much safer bet. Always check out a mirror lens. Sometimes it relies on a glass relay-lens system to achieve close focus for terrestrial use. The cheaper optics designed for cameras invariably adopt this hybrid design. What is needed is a proper astronomical telescope of the mirror-only variety. Newtonians working at prime focus and Maksutovs with a thin corrector window are two of the more common options. They have a few problems of their own, however.

All optics go out of critical focus as the temperature changes in the hour or so before second contact. Mirror lenses are particularly prone to this. There is little to be done about it except to try some form of cover. A body cover does little more than slow the process down, as you are still getting a fair amount of solar energy straight down the optics which are normally in closed tubes. So, don't forget a lens cover. As annularity approaches the reverse takes over and the tube cools. Be aware of what is going on and check critical focus at regular intervals, doing it more frequently as the important events get closer, such as sunspot occultations.

One useful dodge is to consider using a stop to make the focusing errors less critical. With some of the cheaper mirror optics, this has a dramatic effect on image contrast as well. Some say it almost makes the cheaper items usable! Be this as it may, it is worth recommending at all times in solar imagery because of this improvement in image focus latitude and contrast improvement, to say nothing of the heating reduction.

An annulus is needed, i.e. a ring with a hole smaller than the main optics entrance. A stepping-down ring used for filters does the trick very neatly at minimal cost.

The trick seems to work in two ways. First, it increases the effective focal ratio, hence the improvement in focus latitude, and partly explains the contrast improvement. The main improvement seems to be a reduction in the effect of the very common edge/rim figure errors in the cheaper optics. In some mirror optics a terrestrial image borders on the just acceptable, quite useless for the precision and contrast needed in solar-limb imagery. The solar limb at its faintest is plenty bright enough for short exposure times with a suitable choice of solar filter.

There is one final check to make on the optics. Make test shots a little out of visual focus either way and see if it improves the image sharpness. If it does, make a note on the lens where actual focus occurs. Then check if there is a change each time you reassemble the set-up. It will use up some film, but better now, and not when the exposures are for real.

The final checks with all optical systems are not immediately obvious. The first is – will it fit into your luggage? Second, is the overall design going to cause problems with camera shake and stability? The longer the focal length chosen, the more care has to be taken to check the system as a whole. This critical discussion applies in all forms of photography, but no more so than with Baily's Beads and prominences in total eclipses.

Camera Shake

The topic is so important that some time is needed to do it justice. Camera shake is by far the most important single cause of degraded images in solar photography. The trouble is that the camera often gets the blame, when the root cause is often much more fundamental. The problem is the same: namely, heavy bits stuck onto something not designed for the job. Hand holding any video is only just acceptable in wide-angle scenes, and just plain silly with focal lengths of interest in detailed solar-limb shots.

Huge lenses on a camera, or vice versa, upset the natural balance. The slightest vibration in the mechanics starts the whole set-up "nodding and rocking". The best quality SLRs will always give a bit of camera shake, not noticed with a normal lens. Even the best camera and perfect balance are subject to wind buffeting which can, and usually does, play havoc. And there are few eclipses with guaranteed wind-free conditions.

Before we can rectify the problem it is necessary to get the fundamentals right, starting first with the tripod.

Tripod

A very sturdy tripod has to be the most essential ingredient in the whole system. Any serious photographer will tell you that money spent on a good tripod or support is often a much better investment than a complicated lens. Effectiveness is the only important criterion – looks count for nothing when critical sharpness is at stake (see Fig. 8.2).

There are considerable economies to be had in constructing a simple but rigid structure customised for the anticipated solar elevation. When the eclipse is over, leave it behind to make way for baggage allowance and souvenirs.

Tripod sturdiness must be checked long before we get to test exposures. Whilst still in the shop, do all the normal things but always grab hold of the "pan and tilt" or polar axis handle and joggle it about. There must be no floppiness or slack in the head and leg joints, and the legs should not flex at all. If the shop assistant objects, go elsewhere; you only get one chance at an eclipse. Astronomical mounts and drives should pass with flying colours, but you never know without trying them. Nearly all our problems arise from badly designed photographic tripods. Only look at "professional" goods where the user's reputation depends on it.

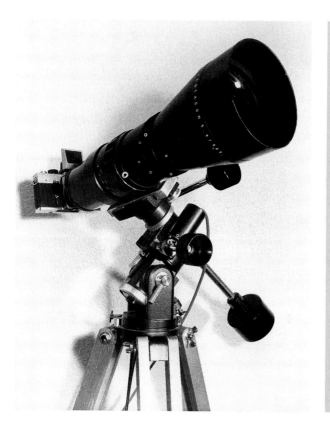

Figure 8.2. A typical tripod and head arrangement. (Michael Maunder)

Provided the tripod itself passes this critical test, the pan and tilt head is next. Anything made from plastics must be rejected outright as being too risky. Apply a modest amount of force to the handle, and reject any with the slightest degree of flexure. It's quite amazing how bad some expensive items are, and how good some of the cheaper ones prove to be.

Checking over the pan and tilt head at the expected elevation is not as trivial as it may sound, as many an eclipse traveller has been caught out by not making that elementary check right at the beginning. Some heads become very "whippy" if not close to horizontal.

The next check is the lens or telescope mount.

Tripod Attachments

Our problems are only just beginning, as simple mechanics takes over.

A professional telescope on a proper mount will be designed well, and the next section is aimed more at hopeful eclipse chasers hoping to adapt some lash-up and expecting it to work. Floppy mounts are the second most common source of grief, after focusing, and possibly the least well understood and discussed.

It is a complete waste of time buying a sturdy tripod if a flimsy head or telescope attachment is added on top. With the chosen tripod and head now in our possession, the next step is easy. Put the whole lot together as it will be on the day. Don't forget the filters or any other gadget, and always load the camera or video with film so that the weights are correct. Then tap the lens with a finger.

No movement or an imperceptible joggle, and you have a rare beast indeed which needs treasuring. You are ready to take pictures.

Invariably, however, movement takes longer than 1/10 s to settle down, and the structure is unusable as it stands. Don't rely on eyesight alone, at the high magnifications we are using it is necessary to be more precise. Check movement with a magnifying glass on a bright spot at each end.

Counter-weights or bracing struts sometimes work. Then try the effect of a hair-drier at close range to simulate strong wind buffeting. The more massive the structure the longer its natural nodding and rocking period about the centre of gravity. The amateur's answer is usually to try to screw the whole lot down tight. This rarely fails at modest focal lengths, but breaks down completely when there is only a single mounting point and a big telescope.

Even if the tripod is satisfactory, camera shake is inevitable with optics stuck on the front of the camera, when the only mounting point is the camera's standard bush in its base. Camera vibrations are transmitted to the lens which has no natural damping except through the head which will always flex in sympathy. That is why it is so important to check its rigidity in the shop. Reject any with obvious "give" or backlash without delay.

Avoid, like the plague, lenses and telescopes without their own mounting bush. Even when one is fitted, and used, the balance point must be exactly through the screw and certainly well within the baseplate of the lens. Adding counter-weights might work, but the problem may be weight above the centre of gravity and the situation worsened. This problem will be detected by our simple tap test.

Whenever camera shake is encountered, always check the natural point of balance of the system. If this is not precisely through the centre of the tripod mount, and as close to the optical axis as possible, forget about sharp images, unless you are very lucky.

Mount the unit at its natural centre of balance, even if this means using a different optical system with a better tripod bush. That still leaves another vibration which cannot be cured by the tripod stability alone. That is sideways whip, often due to floppiness at the camera mount.

One effective solution is the two-point attachment. Mount the optical system on a baseplate, with spacing washers as necessary, to lock the camera and optics together. An accurate centre of balance is also much easier to find on the baseplate which just needs drilling and tapping to fit the tripod head. The secondary advantage of this baseplate is the much larger and more stable contact area offered up to the tripod head.

Cable Release

It is quite amazing how many novices try to work the camera by hand, and wonder about the blurred images later. Cheap cable releases are never an economy and can lead to uneven shutter release if the outer reinforcement stretches or is worn. Sometimes the movement is not quite enough and extra pressure has to exerted to get that final extension to make it work. All in all a source of vibration to the camera which should never be there.

The cable release must be of professional quality and no longer than is necessary to be tied-in out of harm's way with the business end quickly located in the dark. The self-locking variety will be needed later for total eclipses, but that is not necessary now. The essential check is the adjustable extension at the business end so that the camera fires with just enough depression at the hand end. If you have never seen one, get one, it makes all the difference.

Some cameras can only use an electronic cable or have that as an optional extra. Electronic releases are always better than the conventional variety since there is no physical connection to the camera; just an electric cable. The ultimate in this type is the infrared and radio, much favoured by wildlife photographers.

Other cameras do not have a cable release fitting of any sort, and are almost useless for our purpose. It is possible to get around the problem to some extent, but it remains a fudge at the best of times. The self-timer and delay buttons can be used, but these obviously lack spontaneity, and are not to be recommended in the excitement of the moment.

Perhaps the best type of release (IR and radio excepted) has been saved for last – the pneumatic one. As with the electronic, there is no physical contact, but they work out much cheaper and fit all makes of camera with a cable release socket. Most of the pneumatics come supplied with adaptors for the common camera sockets (see Fig. 8.3).

Pneumatics are delightfully simple to work, and there is absolutely no worry in going from one extreme to another. Gentle pressure can be applied until the camera finally shoots, or an excited and frenetic squeeze is quite safe the instant something marvellous is seen. It makes no difference to camera vibration.

It is possible to strap electronic releases together to fire more than one camera at a time, but this risks violating guarantees and damaging expensive cameras if anything goes wrong. Not so with pneumatics. "T", "Y", "X" and other shaped connectors are readily sold in any hardware store, and several cameras can be strapped together. Stereo pairs are an easy option this way.

So far so good, the unit should pass the tap and external vibration tests. The final check is the camera itself.

Figure 8.3. Michael Maunder's stereo pair of cameras triggered by a pneumatic cable release.

Camera

Exposures in the most useful range of 1/125 to 1 s give most bother with vibration. Try some test firings in this range with dummy film in the camera to get the balance right.

Use the magnifying glass test to see if something jumps about after exposure. If so, camera shake is still a real problem.

The only thing left now is the useful technique or concept of a decoupling damper of something flexibly rigid between the camera/lens system and the tripod mount. In everyday language, a thin pad of natural leather or cork can work miracles in stopping vibration. It is amazing how such a simple pad can improve image sharpness out of all recognition.

Provided a rigid tripod is chosen, with a good head, and the optics are balanced and mounted rigidly with decoupling pads, test exposures before and after should show the benefits all round in image sharpness. The effort should be worthwhile. Cameras which "shook" too much before can become quite usable, proving the point that it is not necessarily the camera at fault.

If camera shake persists, dump the camera, at least for this work and get one with less vibration. Any camera with a leaf shutter will never be a problem. Any professional camera with mirror lock facility will score hands down over one without. Its only drawback is the inability to see an image at the moment of exposure.

The answer is simple. Never use the camera view-screen in the first place, except for focus checks and alignment. Use a viewfinder telescope instead, preferably one with an angled eyepiece.

After making sure the equipment sits on the mount comfortably and without vibration, the final check is to see if you can line it up with the Sun at the anticipated altitude and azimuth, and run some proper test exposures.

As with partial eclipses, the safest test object is the Moon, without the solar filter in place. Tests over a period of months will sort out any difficulties at widely varying altitudes.

Type of Camera

Camera choice is usually determined by a budget, and other factors.

Where there is a camera choice, the solar image size has to be brought into the equation. Somewhere is a trade-off with the longest focal length conveniently working within your chosen film or video format, i.e. the image size must be large enough to be useful but not so large that it is difficult to keep within the field of view.

In practical terms this is usually taken to mean a focal length around a metre (1000 mm) for 35 mm, which gives an image size close to 10 mm. This gives about half a solar diameter clearance either side across the width of the film, usually enough to compensate for any backlash in the mounting when brought to the field centre and released.

An equatorial motor drive makes life so much easier, and ought to included in your list of essentials if transport problems can be resolved. For partials and annulars, it can mean a doubling of focal lengths to 2 m or more, and still leave a small margin in a 35 mm frame. The quadrupling of image area can reveal detail quite impossible to resolve before due to halation and other effects.

All is not lost without a drive. A focal length up to a metre then comes into its own if the camera is orientated correctly and the solar image drifts centrally down the long length of a 35 mm frame or video screen. As a first approximation, this image drift takes 2 to 3 minutes whilst still keeping all the solar image in the field of view. For many of the shorter events this is more than enough time to get all the pictures you need between second and third contact. It is certainly long enough for routine solar observation.

The time constraints are considerably eased if a larger format camera can be used. It takes a good 5 minutes for a metre focal length image to drift far out from the centre of a 6×6 cm film frame, once started close to an edge; 6×9 cm and 5×4 inch formats are luxury indeed, doing away completely with drive and guidance problems. Some really impressive focal lengths can also be brought into play.

At the other end of the focal length scale, cameras with standard lenses are particularly good for timed sequences. 35 mm cameras give images at the limit of resolution, and many do not allow multiple exposures. Cameras taking 120 sized film and larger are better for this work. Recording timed sequences are best left to cameras rather than video.

The problems with annular eclipse photography have been covered in some depth because the technology needed is the same whatever type of eclipse up to a total phase. With this essential ground work, the scene is set for the special attractions of a total eclipse.

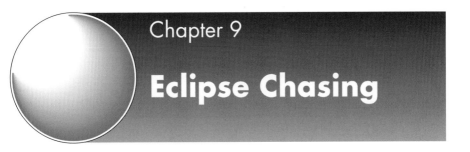

Chapter 9

Eclipse Chasing

Total solar eclipses are such rare and special events that it is really rather silly to be anything other than fully prepared. Bite the bullet, decide to go, then check over some very basic but essential preparations. Here are some ideas to help the process.

State of Mind

Unless there is single-mindedness, some would even call it "bloody-mindedness", problems will discourage all too many, and the experience of a lifetime is missed. There is a famous saying which goes something like – "In later life, one does not regret things done so much as those undone", so keep that factor in mind all the time. Nature waits for nobody.

Planning must start years in advance and the two essential ingredients of time and money are centrally placed.

Getting time off is not always as easy as it sounds, even for the self-employed. Set the time aside *at least* two whole years in advance, and do not spring your bombshell at the last moment on others who have equally important commitments of their own.

Finding the necessary finance is much easier to tackle, as the decision whether or not to go is the critical one, and it should then come down to hoping no family disaster intervenes to mop up the cash. Early saving for anything has to be good budgeting, and some forward planners take out an insurance policy or one of the many similar schemes about. Others see their bank manager. Somewhere in between is the scheme for you. The next critical check is to see if you are chasing the right eclipse.

Time of Year

Unless there is enough cash available, chasing each and every eclipse is impractical, and some serious decisions are necessary. World travel is well organised, but time of year is important in some countries.

Anyone with a fear of insects might find a temperate climate impossible to tolerate. For example, the July 1990 eclipse in Finland and parts of Siberia was well into the midge season. It proved to be no fun for anyone with severe insect bite allergy, even though a site near to the Arctic Circle ought to have been free of all such things according to the naïve. Check your requirements against the lists in the Appendices, then select a country.

Which Country?

Moving along the eclipse track can make all the difference to local weather and "bug" problems. Base a decision, if there are several options, on having a good holiday in a country you would like to visit sometime, or explore further. The reason is obvious; if the weather does take a turn for the worse, at least the rest of the holiday should be memorable.

Some choices are made on very basic issues like, "Are the hospital facilities adequate in case of emergency for known medical conditions?". Other people may have a vaccine allergy, for instance, to the egg-based yellow fever vaccine, and so cannot visit countries forbidding entry without a valid certificate.

Part of the folklore of eclipse chasing always includes true stories of city dwellers totally unprepared to make do with the accommodation conditions, even for a single night, who then give in just before totality.

Everyone wants good skies, but some part of the choice depends on what you want to do. Photographers have to consider the ground on which they will work – paddy-fields or ice – or the type of backdrop, for instance, whereas others are keen on studying the effects on wildlife or plants.

The delicate question of politics and ethics can also arise. The old saying "When in Rome…" cannot be overstated. Anyone harbouring strong objections to a régime or conditions in a country is well advised to make their protest at some later time and place, and not ruin these special events for everyone else by starting a riot. A friendly local bureaucracy can be vital to simply getting there in the first place.

Getting There

The luckiest viewers stay at home for their first eclipse, whilst the rest of us use a specialist travel agent. Only experienced travellers should consider the freelance expedition option, and then only when the country is already well known, or the travel experience has a higher priority. In many countries, all hotel accommodation is snapped up early by the professionals who also have more influence selecting and then varying the viewing site. This ability to switch sites can be crucial, the key being a friendly police force. Wonders have been worked on many occasions in securing last-minute travel permits or supplies. It is interesting to note that after an emergency many freelancers are found near the official groups.

Eclipse tracks mainly over the sea – and there are many in the next few years – mean little option but to take whatever ship is on offer. Ship travel is also the option advised for those with personal aversions to certain countries.

The overriding criterion in any expedition is to be prepared to rough it at the last minute if the weather pattern changes. Weather pattern changes do seem to be more common world wide in recent years, and many an eclipse viewing site has had to move considerable distances at the last moment.

Viewing Aids

Naked-eye views of a total eclipse are really the whole point of going. The sensible idea is to build up dark adaptation so that every scrap of totality is seen to best advantage. Many "trippers" take eyepatches, or the things handed out on aircraft for sleeping to get some dark adaptation as totality approaches. Almost any idea like that, including high density sunglasses, is well worth the effort, particularly if your interest is in studying the outer corona and its delicate tracery.

Totality needs no other preparation, whilst the partial phases just mean making or buying filters, covered in detail in Chapter 3.

No total eclipse is ever truly complete without some form of optical aid, and binoculars must be the most popular, as both eyes are brought into play for a stereoscopic effect. A telescope often has a higher magnification but only for one eye. (Only a raw beginner closes the "spare" and squints with difficulty with the main one.) Having said that, the authors can vouch for the beauty of the higher telescopic magnifications in a schizophrenic stereo view with the real thing in the free eye, something a pair of binoculars finds difficult to do.

Binoculars are quite personal, and never an easy purchase the first time round even for this special occasion. It is much better to make a choice based on some for general use the rest of the time, and a modern lightweight type is unlikely to be a disaster.

Always insist on a thorough test before leaving the shop, preferably with someone knowledgeable. Checking for optical quality is so obvious that it should not need to be mentioned, but that is often the first casualty in a quick purchase. Other essential personal tests include slinging them round your neck for weight, how easy they are to grab in the dark (eyes closed), and can others in your party do the same. Try to afford one pair per person, because short eclipses waste valuable time exchanging a single pair. Provided that the homework has been done, some idea of the elevation will be known. Can you hold them steady for long enough? If not go back to lighter and more manageable types.

The final viewing items are good quality clothing, shoes and seats. Unless the observer is really comfortable, optical aids cannot be held steady and the edge is taken off the pleasure, often due to sheer frustration. Eclipse expeditions always involve a lot of standing around, and so comfortable clothing and shoes appropriate to the climate are a must.

Portable seats are a desirable luxury, many would say an essential, with straw mats a much lighter or disposable alternative. The authors have used and recommend disposable gardening kneeling pads and similar items as an aid to getting comfortably under telescopes and other equipment. Volcanic lava lacerates knees and clothing if this simple precaution is not thought out in advance. A degree of eccentricity is allowable, and a shooting stick is not that offbeat. Zenith eclipses, as in Mexico in 1991, would also greatly benefit from a ground sheet.

A wide-brimmed hat in hot climates adds the crowning touch, and reduces glare in colder climes. The very real hazards, and other special needs in extreme climates are reviewed in more detail later.

Choice of Photographic Equipment

It is decision time. A general theme flowing through the wider discussion for each type of eclipse is the large amount of standing about. Filling that time and recording the rest of the holiday trip might not always be compatible with the best photographic equipment locked up in major experiments. Dedicate equal thought to all phases of the trip.

The fun cameras are not as stupid an idea as they first appear, the more so for video users with severely limited battery power on safari without recharge facility. Fun cameras are a simple answer. They save limited battery power for situations where video scores best, whilst recording better quality still images which can be edited into a more professional travelogue back at homebase, with a small tape recorder for sound effects. The side benefits will be noticed immediately by your friends and relatives who will be spared lots of hackneyed zooms and panned shots! Do these on the higher resolution stills.

Stockpiling film and video tape is a major problem, and best handled by ordering from a reputable and reliable professional dealer for delivery in the week before leaving. Amateur film is excellent and cheaper than the professional stuff, but more variable in properties if very precise exposures are necessary. All the normal photographic films can be frozen for long periods until a day or so before leaving, if spreading the expenditure is important.

Both tape and film should stand up to multiple X-ray monitoring during the personal security checks in most countries, but it is not so much the X-rays which upset tapes as strong magnetic fields. Travel routes should be carefully thought out if there are any doubts on that score. International terrorism is still with us, and hand searching seems to be a thing of the past.

N.B. Cargo luggage X-ray machinery is intended for use on hefty items, and its intensity does not have to be screened for nearby passengers. **Never** commit valuable film or tape to cargo luggage. If the X-rays don't get them, the rapid changes in pressure, humidity and temperature almost certainly will.

Golden Rules of Eclipse Chasing

The Golden Rules of eclipse chasing boil down to five **don'ts**:

1. *Don't* be too ambitious. Take photographic and other equipment, but be prepared to get nothing from it. Rely on the "Mark 1" eyeball, with binoculars a bonus.

2. *Don't* change an established system. An eclipse is not the time and place to read the instruction manual for your latest piece of equipment. "System" also means procedures. To err, is human …

3. *Don't* panic. No explanation needed. See Rule 1!

4. *Don't* lend anything to anyone. Lend the lot after the eclipse by all means. Before that, never.

5. *Don't* let your film or tapes out of your custody. Murphy's Law guarantees that originals get lost or are returned damaged. Make copies specially for publication, loan or projection.

The Aftermath

All eclipse expeditions celebrate their success with spontaneous ideas immediately after the event, and some plan for it in some detail. Give it some thought before leaving home and take something special, or plan to buy on the way, adding that to the overall budget. Experienced eclipse chasers say the celebra-

Figure 9.1. The Aftermath! Left to right: Peter Cattermole, Patrick, Michael, Wendy Maunder and Paul Doherty. Arizona, 1994. *(Michael Maunder)*

tions are the best bit, so try not to miss out on this important – if often overlooked – feature (see Fig. 9.1).

Extreme Climate Trips

Many special challenges face the intrepid eclipse chaser in extreme climates, not least health. Common sense dictates that anyone suffering from an extreme climate-related malady must consult expert medical opinion first.

Because very cold conditions pose more problems, these will be addressed first, then hot and high altitudes.

Cold Climates

Cold climates are best defined as an average daytime air temperature of –20° to well below –30°C. These low figures are unfamiliar to many readers and are best imagined as standing in a domestic deep-freeze all day.

1 *Medical Problems*

The British Antarctic Survey has decades of experience to draw on and they highlight some important health checks which might be overlooked. Top of the list is a full check on teeth fillings. Icy air can have devastating effects on minor tooth cracks which give only intermittent trouble at normal temperatures. Anyone suffering from bad fillings or tooth cracks must seek a dentist's advice, and also take an aspirin supply as a back-up.

Even experienced eclipse chasers tend to forget just how much hanging about there is once equipment is set up and checked. Frostbite is very nasty, requiring expert hospital treatment, and hypothermia is worse. The last thing you need as a souvenir is an amputation or, more serious, permanent illness.

The body takes at least two weeks to adjust its metabolism and other cold-defence mechanisms. Look out for frostbite symptoms which are most likely at the extremities in the hands and feet, and the nose and ears. Keep these well covered at all times and check for any lack of sensation. Confirmation is best left to experienced first-aiders, and not attempted in the open air. The skin loses all colour, difficult to distinguish from a normal cold pallor.

Hypothermia is a very real problem. A characteristic danger signal is when the environment stops feeling depressingly cold and begins to feel nice, warm and cosy. Another early symptom is when shivering stops. Such abnormal reactions must be taken seriously as they can also be present in frostbite. Constant vigilance is the only prevention. Never "do your own thing" and disappear into the vast wastes to get away from the herd. The consequences can be dire.

Hypothermia is a strong possibility for visual observers after the eclipse when the adrenaline dissipates, and there is nothing more to do. Avoid the temptation of alcoholic drinks until well back at base, as they actually accelerate heat loss. Stick to hot cocoa or a similar beverage.

A second surge of adrenaline can be expected during the clearing-up phase, particularly if the majority in the group (naturally) want to get back to civilisation as soon as possible. This possibility must be taken into serious account by anyone with problems when working under pressure. Incipient hypothermia can be missed in the panic of clearing up, returning immediately afterwards in a full-blown collapse. Don't take short cuts like removing gloves and be more methodical than usual, working to a routine worked out in advance.

Do not be put off by the above; it is not intended to be alarmist. Stick together and keep warm is what it boils down to in practice.

2 *Clothing*

Specialist advice from the British Antarctic Survey refers to "loose" clothing, meaning many thin layers of natural fibres like cotton and wool rather than a few thick ones. Boots must be designed for the job, with plenty of spare socks, which must also fit well. Another easily overlooked item recommended is something insulating to stand on. For the 1994 Chilean eclipse a gardener's kneeling pad was ideal for its intended purpose, to insulate the knees when attending to low-angle photography.

The chill factor in winds is often more devastating than the sheer cold. Skiing instructors have a wealth of experience to call on, as do motorcyclists with their long periods of inactivity in special freezer suits. "Long-Johns" are still supplied through specialist outlets such as Damart, and thermal blankets used after marathons are a cheap first-aid accessory.

Hand protection whilst keeping enough flexibility to assemble delicate apparatus, and then to operate it, is given great emphasis by the British Antarctic Survey. Bare skin can freeze to metal fixtures especially when skin is damp, something not impossible when in a hurry and sweating. Surgical removal is not a pleasant prospect to contemplate!

Mittens are one answer, with silk (or cotton) liners for delicate manipulations, returning to thicker gloves for normal wear. Silk, artificial silk or cotton gloves can be bought from photographic or surgical shops. Muffs have long gone out of fashion, and might be difficult to find, unless someone can be found to make them specially. Muffs have a number of practical advantages in that a frozen hand can be inserted extremely quickly, and if this is kept reasonably warm the rest of the time with a flask of hot drink, muffs can serve a dual purpose.

Practice is our recurrent theme. Your health is at stake, not just the results. Practise before leaving home with the preferred hand cover and keep at it until the job can be done in the dark. An eclipse is a dark occasion, when everyone is more fumble-fisted. Check over all the fiddly adjustment screws and knobs.

Hats are important, since a very high percentage of the body's heat is lost through the head. Avoid those with a peak if any viewing or focusing device is to be used, as nudging delicate optical equipment out of alignment is all too easy when a sense of touch is deadened by a combination of thick clothing and sub-zero temperatures. Balaclava helmets are very suitable, as they cover much of the face. Below –30°C, a great deal more care must be taken to cover the face with a balaclava which gives some nose protection. Keep in mind that breath will freeze and cause all manner of other problems.

Specialist face creams, lip salves and hand creams are sold for ultra-low temperature conditions. The risk of getting grease on vital optical equipment has to be balanced against a proneness to frostbite/chilblains.

Assume that the ground will be snow covered and take a good pair of sunglasses to assist in setting up and clearing away. Sun glare off fresh snow is much more of a problem than tropical lighting from sand. Ski goggles are made for the job in hand and will have side pieces to prevent scattered light entry.

3 Tripods

The slightest body shiver will take the edge off your pictures, whether photographic or visual. Hand-holding a heavy video or binoculars is no longer an option in cold conditions. Tripods must have a quick release system, as fiddling around with screw threads with thick gloves is not realistic even for those used to the conditions.

Local sightseers are a problem at any eclipse, and Murphy's Law dictates that someone will slip on ice on to carefully erected apparatus. Be selfish, and chase everyone away. Conversely, non-photographers must observe the etiquette of the occasion and keep a minimum of one body's length away from any equipment.

Ice and permafrost are real problems for tripod stability when rubber feet go rock hard, slipping just as easily as metal. Give deep and serious thought to the problem of anchorage. Ice around the feet can be melted and re-frozen for a perfectly rigid ground contact, but bear in mind that gas devices are forbidden on aircraft.

One author's answer to the sturdy mount problem was used with great success in Montana in 1979. A "G" clamp camera mount with a wood screw fixture at the bottom is normally screwed into a post or tree, allowing the camera to be pointed in any direction. On compacted snow the clamp is simply screwed into the snow to make a perfectly rigid fixture, more than adequate to support a Lubitel for sequence photography.

4 Telescope Type

Stopping down mirror lenses is considered in more detail later, as focus shifts are a common problem in the rapid temperature changes at all eclipses.

Ordinary glass lenses are a serious option in cold climates, in spite of their extra weight and bulk. Infinity focus is less critical to find and is often the last stop in the adjustment, needing no thought at all. Mirror lenses were once the only choice for critical focus at all wavelengths. However, this restriction is becoming less valid with the wider use of low dispersion glasses and aspherical optics, and even simple glass optics working at $f16$ or longer may perform better under rugged conditions.

The final detail to check with any lens is how much easier it is to focus at low temperatures. Any reputable lens will be manufactured with greases designed for the job. This is not true for all non-marque lenses, or good ones given years of hard work, picking up grit. A simple test method is outlined later.

5 Cameras

All camera lenses and viewfinders suffer from dewing problems as the air warms up and then cools towards totality. In cold climates frost rather than water forms, which cannot be wiped off without irreparable damage to the surface coatings (see pages 125 and 126 for a discussion of the special attention needed after the eclipse). Videos are usually fully sealed and ruggedly designed for all-weather conditions.

SLR cameras are a particular problem, because of the larger number of moving parts compared with a TLR or rangefinder. Mirror vibration can increase badly at very low temperatures due to stiffening hydraulic or spongy dampers, or lubricant thickening. Russian cameras were designed with harsh winters in mind with the right sort of lubricants. Consult a qualified repair company before changing any lubricants in non-Russian equipment.

Autofocus lenses and any other device relying on mechanical movements initiated by the camera itself should be regarded as taboo. Stick to any mechanical one, not using a battery of any sort (except for normal metering checks), of the type sold for professional use. The Bosnian war revealed many failings in modern cameras, whilst those built prior to autofocus stood up to the cold conditions very well.

Autofocus consumes power, and motor drives considerably more, leaving little over for electronic controls which behave erratically at the anticipated temperatures. Consult the instruction book or manufacturer for advice on a minimum working temperature. Most cameras should work, but find out now, not on the day.

The Cinderella of cameras these days is the ciné. There are both Super 8 and 16 mm cameras from Russia almost tailor-made for the conditions, doubly interesting because they do not rely on battery power at all, being clockwork powered.

"Read, then follow the instructions" is the best and only advice to be followed with any equipment, including any remote controls offered with the kit.

6 *Tapes*

Video cameras do not use photographic film, but the recording medium is usually backed on to a similar plastic material which will also become brittle at low temperatures.

There is a very good reason why only the best recording medium should be used at ultra-low temperatures. Silver-halide films are prone to scratching and video tapes are not immune to degraded images due to head contamination with more material than normal picked up by static or any other obscure effect like metallic contraction. The result is the familiar "snow storm" effect on play back and copying at normal temperatures. Whilst the effect is unlikely it is not impossible, and may even be transitory when temperatures return to normal.

Good quality equipment should be proofed against excessive head contamination, but with so much new equipment on the market, not necessarily intended for such low temperatures, be aware of the risks. Prevention is better than cure and extra care is necessary to give all recording heads a good overhaul and clean before leaving. Time will tell how recorders based solely on digital memories perform.

7 Silver Halide Films

Films are now of such a high quality that it is true that none readily available in the modern world will be substandard, and any choice is entirely dependent on the personal style or type of photography. Film sold *en route* could well be a different story with no guarantee of good storage, or locally-manufactured materials. As with video, take plenty of stock.

The cold requires attention to a number of factors in order to get the best results. First, and foremost is brittleness of the acetate base. Take extreme care at the end of each reel not to wind on too energetically because it is not unknown for the film to snap and part from the spool. At –30°C and below film has also snapped in the middle of a reel.

Do not rely either on the clutch in motor-drive cameras which is designed to work at normal room temperatures, if there is a surge in motor power. Better still, avoid motor drives to conserve battery power.

Another film brittleness effect is in the gelatine emulsion, which becomes rock hard at low temperatures. Common sense logic dictates that a rock hard substance should not scratch easily, but it does. Grit also becomes tougher, particularly if it contains moisture, and the felt cassette lips also harden. The nearest analogy is to describe film as "being subjected to abrasion between a rock and a hard place".

Our problems do not end there. Brittle film is extremely difficult to load into cameras. Strange arcs and circles appearing on exposed frames are usually caused by pressure marks, and "rucking" in the frustration of trying to jam the stuff into the guides. It is surprisingly common at room temperatures, particularly with 120 sized film which is difficult to load into developing tank reels.

Some films are coated on Estar bases and these must be treated with even more care. The base is so tough that vital parts of the camera will break long before the film will snap. Motor drives must be perfect and extreme care taken in winding film on by hand if the fingers are numb and there is any doubt over the amount of force being applied. Many a camera has been wrecked by extreme force applied on an Estar-based film.

Another reason why motor drives should be treated with care is due to their very speed of operation, ideal conditions for static electricity. Contrary to lay expectations, the humidity is not zero as ice is one of those rare materials which sublimes directly from the solid. This is why dewing is always possible in sub-zero temperatures. As film is moved fast through the felt lips of the cassette, the static electric charge can be impressive. It can be seen quite clearly as bright sparks in a cold darkroom when film is removed from the cassette. The effect on a film is devastating. When processed it will show bright patches and in extreme cases with the faster films, an almost total fogging. Sometimes the image will be reversed.

This is known as the Clayden Effect, and is caused by light of moderate intensity on top of a latent image produced by a very brief exposure, such as a spark. The Clayden Effect is not as improbable as it might seem at first. Strange black marks and reversals might just be due to it. Low-intensity, long-exposure corona pictures are perfect conditions to expect the Clayden Effect with partially reversed images. There have been no reported cases of static electricity upsetting camera electronics up to the time of writing.

Moral! do not use motor drives (battery power drops catastrophically anyway) but inch film on very gently … and that goes for 120/220 workers, too.

Some modern cameras rely on the film being unwound from the cassette onto a spool before exposures start. Any camera with this feature should be loaded at normal room temperature, long before taking into the field.

The final point to watch with most films is the question of film speed. The classic way to improve reciprocity failure is to cool film. The effects are not likely to be important at short exposures, but can lead to some interesting possibilities with extended coronal exposures. Consult the manufacturer's literature or subject film to the test method on page 126.

8 Battery Power

It is an unfortunate fact of life that many of the more common batteries fail catastrophically at quite modest temperature drops. Normal carbon/zinc batteries do this, becoming almost useless below freezing point. Alkaline batteries fare little better, but nickel–cadmiums (NiCads) retain some usable capacity down to below –10°C.

It is a case of be guided by the camera maker. Canon were the first to market what is now a common autofocus lens system in their T-80. The instruction manual specifically forbids NiCads, which means that such a system almost becomes so much luggage ballast in freezing conditions.

Another camera of the same vintage, the T-70, has no such restrictions on NiCads, and uses what are, in camera terms, huge AA cells, compared with the T-80s AAA size. The efficient motor drive shows promise of being usable at our intended temperatures. The instructions, however, draw the line well before this point.

These figures, drawn from the instruction manuals, are very informative. What they mean in practice is that there is no guarantee of exposing a single reel of film even with fresh carbon–zinc batteries at the modest temperature of –10°C. The diminutive alkaline AAA cells of the T-80 give the approximate capacity of the lithiums used in many of the modern cameras following on from it.

Video cameras are at a distinct advantage at low temperatures because of their dependence on NiCad rechargeables, the only common and cheap camera batteries retaining good power characteristics at low temperatures. Normal cameras often forbid NiCads because of their very low internal resistance. If anything goes wrong, current supply is almost unlimited, risking complete burn-out of the electronics or wiring.

Much more likely is a slower motor drive action due to film stiffening and lubricant thickening. At slow motor speeds, current is limited by conventional battery internal resistance, but not so with NiCads and many of the newest lithium cells. Current demand leaps, depleting an already limited capacity. Forget motor drives unless they function well enough under test conditions, or only a few exposures are needed.

Ruggedised lead acid batteries used in vehicles stand up to the conditions reasonably well. Unfortunately, the voltage rarely matches camera needs, but is often specified for telescope polar drives. There are some sealed units designed as portable power supplies and where aircraft weight restrictions allow, choose such batteries for high capacity work. Do bear in mind, though, that all batteries must be declared and taken as hand luggage on planes. X-ray scanners are programmed to detect batteries used in detonators, or lead, as in bullets. Always have a back-up power concept in mind in case the goods are seized. You have been warned.

The conclusion to be drawn is that secondary cells such as lead acid and NiCads are to be preferred because of their superior low temperature characteristics and high current delivery. Rechargeable lithium cells and similar "exotics", or the newer primary cells do not have a sufficiently well defined track record to risk yet, but it will happen.

All battery-dependent devices must be assessed, including tape recorders, radios and any other observing accessories needed on the trip. The more essential devices like pacemakers and hearing aids must be given special attention to ensure that they are well protected by clothing, with plenty of body heat. It is surprising how cold the ears can become, and whilst not as cold as the surrounding air, be sufficiently cool to lose power during a long observing session. The batteries are simply too fiddly to replace in the open air, and are difficult to manipulate for correct polarity in low-lighting.

There are many proprietary devices sold to keep batteries warm, and many require the batteries to be kept close to the body. All require extension leads with the consequent risk of entanglement and poor contacts. Some cameras are deliberately designed to prevent the use of external batteries. All kits requiring something other than the camera maker's intended battery usage, must be checked over and then double checked by the test method described here. It is much better to concentrate on a different type of camera which does not need battery power. Russian cameras are the obvious choice here.

9 *Plastic Items*

Strictly speaking, plastic items are items which have plastic properties. Some are truly flexible, others almost metallic in rigidity, particularly when reinforced with glass or other fibres. All plastics become brittle at extremely low temperatures. and can shatter when subjected to quite modest forces or shocks.

Pay particular attention to any electrical leads where the insulating material might fracture in a knock which is quite harmless at normal temperatures. Subject all vital leads to the test régime. Even if the material does not fracture, delicate photography might be upset by vibrations due to wind buffeting from the rigid leads.

All plastic fittings must be treated with extreme care and eliminated wherever possible. Specific points to concentrate on are any load-bearing structures, such as tripod fixtures – notorious weak spots. Some cheap tripods have entire tops made of crude moulding, and many of the more expensive light-weight varieties have the entire leg structure in a single mould. Use only those guaranteed to withstand the expected temperatures.

Discard any modern lenses constructed from plastic bodies rather than metal. If excessive stress is placed on the body because of stiff focusing, from whatever source, damage may result. The manufacturer should be able to specify any temperature limits.

Perhaps a more serious consideration is the wider use of plastics in camera bodies. Unless the lens is well supported, and we are likely to be considering some heavy optics for the long focal lengths, distortion or fracture is a distinct possibility. Weak points are not confined to the lens mount, but can include the tripod bush area which is metal but set into a plastic body. Never over-tighten this, reviewing the whole design for balance, and do not allow heavy lenses to be unsupported on the front of the camera. Should a fracture occur, the chances are that the whole film will be ruined by exposure to daylight, and the camera itself an expensive repair, or write-off.

Select freezer quality bags (which should not go brittle) to store all equipment after the eclipse for the essential aftercare.

10 Aftercare

All vital pieces of equipment must be treated with just as much care after the eclipse as before to avoid condensation. By far the simplest method used over the years is to carry resealable polythene bags for each piece of equipment. As each piece of experiment is dismantled, the item is returned to its dedicated bag and sealed in, to be opened only when the whole has reached room temperature back at base.

Films need special care. It is always better to leave films in the camera and seal the whole lot, rather than risk film breakage on rewinding with all the attendant static electricity displays. Motor-driven film rewinding is only a last resort.

Batteries ought to be removed from cameras if the camera cannot be sealed to stop shorting if condensation forms on rapid return to domestic humidities.

11 Test Method

It will be clear by now that many of the cold-weather problems can be eliminated by the careful choice of equipment and battery type, but the worry still remains whether or not it will work on the day. The only logical way to check this is to travel to a similar climate and try it. However, this is not a practical proposition for most, and the following test method has been found to be an excellent sorting test.

If the equipment fails the test, it will certainly not work on the day, whereas the converse of passing and working later cannot be guaranteed but it is a better check than nothing.

Requirements: A domestic deep-freeze, preferably a chest freezer; large resealable polythene bags.

Method: Do not attempt this with anything valuable or totally dependent on electronics and/or liquid crystal displays. Limit camera tests wherever possible to fully manual types and lenses.

Load each camera with dummy video tape or film, preferably out-of-date from any shop, and check each battery for full charge, then add any lenses and seal everything into a large resealable bag. Place the bag in the freezer and leave to equilibrate for several hours, preferably overnight. Most freezers are preset to $-18°C$.

The next stage is easier with a chest freezer. Whilst still at low temperature, run through a simulated exposure sequence. As each piece of equipment performs according to specification, the experiment can be repeated as often as necessary using the chosen hand protection until fully proficient. If the freezer has a fast freeze or can be adjusted to lower temperatures, proceed only if the first stage is satisfactory. At the end of the experiment allow at least 12 hours for the equipment to warm up to room temperatures before opening the bag.

All lenses must be studied intently for iris stop-down/open-up (whichever system is used on the camera) and for unacceptably sluggish action.

This simple test will establish any weak links, and battery capacity or potential problems with the recording medium being sensitive to static electricity. Get films processed only and examine for abnormal exposure patches due to static or stress marks. The method is the only practical way of testing video cameras for battery capacity and recording head build-up, but must only be attempted if the temperature range is permitted in the instructions.

Hot Climates

Hot climates fall into two types, hot and dry, and hot and humid, but for our purposes we will discuss them together with a definition of an average night-time air temperature above 25°C (77°F). Most countries will achieve these temperatures sometime during the summer.

1 Medical Problems

Because our definition covers a familiar temperature range, there ought to be no difficulty in planning ahead using travel guides, based on familiar temperatures back home. The problems arise when the range goes way above normal, and the humidity follows it.

The primary concern is sunburn. Below a certain angle (determined by the ozone layer concentration) the Sun might not burn at all, and above that point the risk increases alarmingly. It is not realised generally that the burn risk is not a simple relationship to the angle of the Sun's elevation, but to the tangent of that angle. A few degrees difference means that exposure times have to be cut from hours to minutes. Close to the equator, everyone is aware of the dangers, so regard an increase of 5° from home with the same common sense and err on the safe side, taking that modest difference to be the same as spring to summer conditions.

Burning is now closely linked to skin cancer, perhaps the largest single cancer in some populations. With so much standing around at eclipse sites, in cloudless sky conditions, particular care must be taken to ensure that youngsters have adequate sun protection, hats and high factor sunblock. Too often eclipse trips are marred by unthinking parents.

Closely related to sunburn is sunstroke, which like heat-stroke, can be life-threatening. Each party ought to have someone well versed in the symptoms which are not always easy to spot in advance – precipitous collapse is often the first indication. Be on the look out for headaches, "prickly heat" and any abnormal behaviour.

Prevention is the best cure, and apart from sensible clothing, sunblock creams and shade, adequate liquid intake is generally accepted to be the best preventative. At one time a large salt intake was recommended, but there are mixed views on that now. It is clear, however, that the body's overall electrolyte balance is critical and that potassium features highly, rather than sodium.

Do not be tempted to use potassium chloride instead of ordinary salt, as this should only be used under medical supervision; rather drink plenty of natural (whole) fruit juices which contain enough minerals to keep you going. Sports clubs and marathon runners can be consulted for alternative proprietary electrolyte replacement drinks.

Personal hygiene is rarely discussed in travel brochures, and is seldom too much of a problem at low humidities. In high humidities obvious problems do arise, with risks from fungal as well as bacterial infections. Take bactericidal soaps, and where the water is known to be soft, take some sodium bicarbonate as one of the safer bactericides to add when necessary. Attention to vaccination and other local requirements is essential, but some people react adversely to the medication. Allergy to yellow fever vaccine has already been mentioned, and some of the malarial remedies are becoming less effective and cannot be used by many.

The history of eclipse trips is littered with tales of participants having an adverse reaction to some medication or other. Without labouring the point, about 10% of travellers on average leave their vaccinations until the last minute and ruin the trip trying to recover. Attend to this essential, but neglected, point well in advance.[*]

----

* Sweaty hands have been known for a long time as the most common cause of bloomed lens surface, and other delicate surface destruction. Cotton gloves are a wise precaution in the extra stress at eclipses. Sunblocks and other creams such as lip salves are also damaging to all optical surfaces, many types of plastics, and recording media from digital to silver

2 Clothing

There are so many specialist shops dealing in the necessities that little needs to be added other than a recommendation to visit them rather than an ordinary outfitters. Concentrate on natural fibres such as cotton and wool, and leave nylon and other synthetics at home. Wide-brimmed hats are strongly advised.

3 Tripods

The main requirement is stability and good design to stop dust getting in the movements. Normal pointed feet might not be suitable on loose sand, and some alternative such as digging in deep might be necessary, in which case dust access prevention becomes a higher priority.

4 Telescope Type

Optical equipment does not take kindly to high humidity. Lens surfaces are particularly prone to grow fungus, and this can and often does spread to cements and other mounts. Fortunately, eclipse trips rarely involve a long enough stay for the full-blown problem to surface (literally). It is sufficient to be aware of the potential and take avoiding action. Similarly, high humidity can lead to severe corrosion problems when different metals are used in construction. Reputable manufacturers are aware of this, but eclipse trips often mean some basic rules are forgotten. Keep all optical equipment bone dry.

Dust damage to cameras is discussed next. It is accepted common wisdom to always use a haze or UV filter on all lenses, and to keep them there at all times.

Continued

halides. Similarly, insect repellents. Many repellents are based on diethyltoluamide, which will destroy many common plastics, and cause marked colour shifts if the vapours interact with colour film. Avoid the use of nebulisers, insect repellent vaporisers and aerosols when any colour films are out of a secure container. For our purposes, a loaded camera is not a secure container. The effects are minimal in conventional scenes, but the whole point of eclipse recording is the subtle tonal detail and range.

Such filters are infinitely cheaper than damaged optics. If a dust cap is supplied, use that as well in a "belt and braces" philosophy.

5 Cameras

As with tripods the main problem is dust. The majority of videos are well designed and dust access for a sealed unit should never arise. The weak link is the loading door for the cassette or disk, and with much high precision engineering inside, it pays to be careful. By far the simplest option is to seal the doors with masking or reusable insulating tape.

Still cameras are much more of a problem if not of the disposable or sealed type, because of the larger number of interchangeable parts. Some of the older cameras, and modern large format cameras, still use unsealed units for the mechanical cocking shutters. These are unsuitable in dusty conditions, and are easily ruined by the slightest speck.

Never rely on the seals entirely whilst travelling from place to place, and put into large resealable plastics bags instead of the normal carrying bag. The reason is very simple; joggling in transit can expand and contract the bag, very much like a lung, and dust is sucked into the camera works. Many a camera has been ruined in transit this way.

Do not wipe dust off the works, and never from any optical surface, otherwise irreparable scratching is inevitable. Use a photographic brush, preferably a blower to remove the bulk and leave any finer particles held by static well alone, and just put up with the degraded images. They will be much better than the permanent effect of scratched optics! There are a number of specially treated cleaning cloths on the market to be used before setting out, and when back in clean conditions such as an hotel room. Some contain an anti-static preparation, a worthwhile precaution to assist brushing.

High humidity can cause difficulties during long stays, but rarely for short trips. Build-up of fungus and other phenomena takes a while to develop, and the major problem is the simple one of moisture deposition on the vital electronics and lens surfaces. It is the converse of the cold climate problem, and seen most frequently on leaving a freezing air-conditioned hotel room for a tropical environment. Moisture can form within seconds and is next to impossible to eliminate in near 100% humidity. The only solution is complete sealing into plastics bags until the equipment has reached street temperatures.

Moisture formation in optics is not restricted to the lenses, but also to the viewfinder, which is extremely difficult to clean at the best of times. Desiccants can be used, it is really a question of stopping the problem by sensible planning.

The second major problem with all types of camera is due to the heating effect of the Sun. Most cameras are black and absorb heat very quickly. In tropical climates the internal temperature can easily exceed the recommended limit. Both the electronics and the film are then ruined. Even mechanical cameras, empty of film are not exempt, as some lubricants and plastics decay extremely rapidly under these conditions.

It is rare not to find a camera instruction book forbidding the storage of these delicate objects in car glove boxes and the like. Air temperatures can easily exceed 60°C (when egg white hardens), and reactions with petrol and other vapours are accelerated. Contact with some plastics can lead to permanent fusion or other damage.

6 Tapes

Along with disks and other electronic media, dust and moisture are best avoided. The delicate tracery of the corona is difficult to repair when attacked by a snowstorm arising from dust or moisture on the recording head, and as much care is needed to avoid either as in conventional photography.

7 Silver-halide Films

The quickest way to destroy a film is by exposure to high temperatures and humidity, and every attempt must be made to keep these factors to a minimum. Consult the section on low temperatures for the effects of static electricity at low humidity, also common in hot deserts. Similar restrictions on camera motor drives might apply for the same reasons.

The characteristic effect of a high temperature and low humidity environment is an accelerated ageing, called "ripening" in the trade. Some hypering methods rely on this drying effect for part of the process, and once passed through a peak of efficiency, the film "dies" very rapidly indeed. Amateur films are made to cope with as many varied and likely climate conditions as possible, and these are preferred, provided that they are within their expiry date.

Professional films are a totally different proposition, as these have been ripened to their peak efficiency before sale. That is why correct storage at 13°C (55°F) or lower is specified until the last possible moment, and rapid processing. Professional films can only be recommended to those who know what they are doing.

Almost as an aside, the difference between amateur and professional film can be exploited in eclipse photography, specifically for longish coronal exposures. When properly stored in temperate climates, film might (just) have ripened to a point of maximum sensitivity by the expiry date (there is a wide margin to allow for higher temperatures, in the USA for example). It is certainly worth checking out any short-dated stock for improved long exposure response similar to hypering. There are no guarantees, but many professional photographers have used this technique to fine-tune both speed and colour balance response.

Without doubt the real film killer is humidity, and above all, rapid changes in it. Moisture pick-up is essentially prevented when film is in its original container, and films should be left intact until needed. Moisture pick-up before exposure can lead to variable response, particularly if film is not evenly wound and extra moisture collects in the looser areas.

After exposure, the effects of moisture are even more devastating, mainly because the sealed packaging has been opened. The latent image is rapidly destroyed by excessive moisture, and many corona pictures are ruined this way as they involve fainter and more delicate colouring. Pay particular attention to the best methods of stopping rapid temperature changes, and how to seal films after exposure. Remember, desiccants can make the situation worse, and the manufacturer's advice should be followed for the optimum humidity; always obey the instructions for the optimum delay time when opening containers after storage at low(er) temperatures (Table 9.1).

High humidity conditions can also risk fungal and other pest attack on the delicate gelatine emulsion. Prevention is the only cure.

Table 9.1. Warm up time in hours (minimum) for a temperature rise to site

From film type	Air conditioning 10°C (20°F)	Fridge 30°C (55°F)	Deep-freeze 50°C (90°F)
35 mm, 110, 126	1	$1\frac{1}{2}$	3
10 pack	$1\frac{1}{2}$	3	6
35 mm bulk	3	5	12
8 & 16 mm bulk	1	$1\frac{1}{2}$	3
120	$\frac{1}{2}$	1	2
10 pack	1	2	5

APS cassettes have not been around long enough to be fully idiot-proof under all conditions. Conventional 35 mm cassettes certainly need special care in hot dry atmospheres, and dusty conditions. Modern plastic-bodied cameras pick up static electricity very quickly, and the human hand can easily transfer this via the metal transfer mechanisms to the internal workings and then minute dust particles, invisible to the naked eye are attracted to vital sites. Scratching does not necessarily have to be the fault of the cassette or its felt lips, although that is the normal source. Static electricity can be transferred to the felt via the metal cassette body.

Limit handling as much as possible, and keep motor drives to a minimum in ultra low humidity climates when there is a lot of dust about.

See also the footnote on pages 128 and 129 in the medical discussion. Vapours in car glove boxes are now mentioned in most film guides.

8 Battery Power

This is rarely a problem in hot dry climates. Avoid moisture build-up in humid conditions which can short out batteries or electronic circuits. Suggested first-aid action is to remove batteries until needed.

9 Plastic Items

See the footnote on pages 128 and 129. Many plastics also sweat at high temperatures, and the materials exuded are difficult to remove. It is also not unknown for dissimilar plastics to weld when heated. Stick to the more reliable construction materials such as metal and wood wherever possible.

10 Aftercare

Similar considerations of moisture prevention apply as in a cold climate. Dust is the new feature.

11 Test Method

Tests can be tried at any time by simply leaving the equipment in the Sun or in a glasshouse to reach even higher temperatures. As in a cold climate, use mechanical equipment and never risk complex electronics.

Sunshades or parasols must be tested for security under windy conditions; a hair-drier is as good a system as any for this purpose.

High Altitudes

The 1994 eclipse in Chile and Peru was the first experience for many eclipse chasers of high altitude conditions. We define high altitudes as anything over the equivalent of a pressurised aircraft or 2500 metres (7000 feet), or where breathing and related medical problems become important (see Fig. 9.2).

Although a special effort has to be made to find a site at altitude, it is often the best chance of clear skies free from atmospheric pollution, or a low moisture content for special experiments. It is no guarantee, however, of cloudless conditions and recent eclipses have demonstrated that the weather pattern can change or reverse at the last moment.

High altitudes are no more than a special case of cold climate trips, and little extra work is needed to prepare equipment. Altitude-specific problems are medical.

Medical Problems

Whole textbooks are available on the subject, and a full medical check is advisable if more than a few months

Figure 9.2. Mid-eclipse at 14,000 feet in Chile, 1994. Shows Venus, top right. *(Michael Maunder)*

from the last one – youth is no guarantee of immunity. Many of the serious cases seen in Chile and Peru involved younger people, partly because they assumed "fitness" and partly because metabolic rates have some bearing. Perhaps, also, older people were wiser and took things easier.

At anything in excess of 4000 metres (1300 feet) altitude sickness can strike anyone at any time. Experienced eclipse chasers can be struck down for no obvious reason while the novices are unaffected. Prediction is extremely difficult, and the 1994 experience seems to indicate that a form of mass hysteria or suggestiveness plays some part as the adrenaline ebbs and flows.

The long periods of "hanging about" do seem to have some influence, as these give plenty of opportunity to be introspective. On the other hand, a particularly good eclipse, such as in 1994, burns up a lot of energy, and the excitement on top of a lack of sleep and the anticlimax afterwards caused many unexpected difficulties. Mountaineering groups have much advice to hand on and leaflets are now available from all the major observatories, with a disclaimer, of course. The salient features seem to be:

1. Take it easy reaching altitude. Observatories recommend at least one staging post on the way up for an overnight stay. Almost a quarter of the 1994 groups found the transition from sea level to 4000+ metres too much in one go, and they had to move downhill.

2. Take it easy at altitude. Rapid movements unless acclimatised can lead to permanent injury.

3. Avoid alcohol completely until after the event and back in familiar territory. Migraine and other illnesses affected people for the first time when they ignored this simple rule.

4. Maintain an adequate level of liquid intake. The low humidity at altitude is less obvious than in a hot desert. Many medical difficulties have been traced back to excessive water loss due to unseen sweating in the excitement of the moment. Fresh fruit juice is a good idea for electrolyte control.

5. Work to a set plan. Altitude confusion is insidious and creeps up on us all. Plan to take two or three times longer than normal to complete tasks. It is also a good idea to work in groups to check on each other.

6. Keep the plan and equipment to a minimum, to save physical effort moving stuff about. The observing site could be a long way from the transport.

7. Visual workers will find a diminution of visual acuity, particularly colour sensitivity. Pilots undergoing altitude training describe a shot of oxygen as being like "switching the light on", as good a description as any of the phenomenon. Binoculars or telescopes should be relatively fast to compensate, sacrificing magnification if necessary.

There are drugs available for altitude sickness. They cannot be recommended for self-administration unless prescribed by a doctor familiar with all your medical history.

Chapter 10

Total Eclipses

The die is cast, and preparations begin in earnest for the next total eclipse and all our hard work practising with daily observations and at partial and annular eclipses starts to pay off. Lessons discussed in earlier chapters reach their culmination in the very special atmosphere surrounding these rare events.

Apart from saving hard – some very basic decisions need to be made right now.

Photography or Viewing?

The real question is not so much "Which"? but "How much of each"? Concentrating solely on photography is a huge mistake, as is standing gazing when there is suitable equipment ready and waiting to be used.

There is an old adage in photography, which has its mirror in most activities, which is best summed up as the "KISS" principle: Keep It Simple, Stupid!

Anyone on their first trip, and many others for that matter, will do well to keep that essential message uppermost in their mind as they proceed. Plan by all means, but always have a simple fall-back position which boils down to doing nothing but having a good time.

At least half the time ought to be devoted to visual work. Provided luggage and budget restrictions allow, an active photographic plan can be combined with

Figure 10.1. The chaos of an eclipse site. Java, 1983. (Michael Maunder)

binocular or telescopic views. The essential thing is a firm base to hold all the equipment and some form of visual aid such as binoculars substituted for the camera viewfinders, always a good plan with their superior quality over a ground glass screen or camcorder monitor.

Practice is then essential to carry out the camera exposures by rote and feel, and to know how to collimate everything so that all bits of apparatus point in the right direction, every time, and well in advance of the start of the phenomena. There is nothing worse than a last minute hunt for a hugely magnified image on to a small screen, the more so if travel arrangements (say due to weather) mean hurried arrival on site and a huge scramble for the best positions (see Fig. 10.1). One author (MM) missed getting any telephoto pictures for this very reason due to traffic delays in the snow. Late arrival on site in Mongolia in 1997 meant that only hurried wide-angle shots were possible.

Video/Camcorder

In earlier chapters, the superiority of these devices in recording the subtle pinhole effects under trees has been noted. They are also the instruments of choice to record shadow bands in real time and related phenomena, and all other transient lighting effects across the scenery. The "roller-ball" effect as the Moon's shadow passes by is always most impressive.

Electronic devices tend to fail at the most critical moments of Baily's Beads and the Diamond Ring effect, because the safety cut-outs are not designed for such intense lighting in a small area.

Because of the uniqueness of the event, there is very little testing that can be done, and it is best to trust to luck that the sensor or electronics will not be blown in the process. The final result usually displays a distorted star effect across the whole frame as pixel lines overload in turn. Some photographers like this, others detest it, but one thing is certain – it is extremely difficult or tedious to clean up afterwards, and best left intact.

Ciné

Many photographers are obsessed with the newer technologies and forget that ciné has many advantages into the foreseeable future. The most obvious ones are a superior image resolution, cheapness on the second-hand market (making them almost disposable!), and a huge range of excellent optics and adaptors to telescopes.

The main disadvantage is that, at an amateur level, ciné photography is obsolete as far as dealers, suppliers and processing laboratories are concerned. It is very difficult to purchase even Super-8 film now, or 16 mm in small quantities.

"Fun" Cameras

The virtues of these cameras have been discussed in earlier chapters. Although the purist will blanch at the suggestion, any simple camera has to be lumped into the fun category on the KISS principle. Minox to Rollei, it matters not, simplicity of operation and instant readiness do.

SLRs

The level of complexity of single-lens reflex cameras has increased enormously in recent years, but so has the potential for creative results. Carrying one or two extra bodies as a back-up is always a good plan, particularly if these are mechanical rather than electronically controlled. The second-hand markets are a reliable source of cheap goods which can be checked over for reliability long before needed in earnest.

TLRs

Twin-lens reflex cameras are very much under-rated, although they are no longer manufactured except in the Far East. Nearly all have superb optics, better than 35 mm, and most have leaf shutters completely devoid of camera shake. The simplest of all is the Lubitel and every eclipse traveller should carry at least one of these, which cost little more than most 36 exposure films with processing. Normal picture quality even in the latest versions is much better than given credit in the press, and the authors have yet to find a simpler answer for timed sequences. A pair of Lubitels has followed the authors around the world since 1976. With a little ingenuity, they can also be used in afocal mode, and in stereo pairs (see Fig. 8.3).

Rolleis came back on the market for a short while and Mamiyas, Yashicas and similar makes can still be picked up second hand in pristine condition. The main source of new cameras similar to the Rollei layout is China. The Mamiyas provide a particularly interesting option with interchangeable lenses.

Larger Format

The practical advantages of superior optics and image quality are legendary. It is assumed that anyone taking this option has already discovered those facts for themselves and know what they are doing. Always use the largest format possible.

Ccds

These have been left till last because using them does seem to be unnecessary. Normal video and camcorder technology is a more developed technology and a better ruggedness to cope with lighting extremes.

Even the outer corona is quite bright in terms of the faintest image a CCD can detect, and risking such expensive pieces of technology to Baily's Beads is not something to be taken lightly. There is also the question of the cost of units with an area even remotely approaching 35 mm film, but that is changing rapidly in professional circles and will eventually filter down to the amateur.

CCDs and allied techniques will come into their own, and increasingly so in the next few years, because of the very same sensitivity which limits them as an alternative to the more familiar technologies. That is in spectroscopy and similar studies, where the faint light is split up into its component regions, particularly in the near infrared, and some of the more interesting spectral lines can be studied in greater detail.

The future of CCDs promises to be an exciting and assured one, but in the more esoteric science rather than pictorial rendition.

Lenses

A bewildering array of lenses faces the uninitiated, and whilst many types have very specialised applications, we are only interested in the common ones. Some of the basic properties such as field of view and transit times are important in planning ahead (see Appendices D and E).

Standard

The standard lens is the workhorse in all photography, and so it is at eclipse sites. Optical quality is at the highest and lens speed usually a stop or two faster than most other focal lengths.

Apart from these practical advantages, a standard lens is intended to generate an image as close as possible to that seen with the naked eye. Use a standard lens whenever bright ideas run out.

Wide-Angle

A wide-angle lens obviously has no problem in finding the centre of interest, and focus is so much easier.

Depth of field even at full aperture is impressive, and this can be terribly important in the heat of the moment when time is critical, and becomes even more so when light levels drop too low to see well enough. Focusing can rely on the depth of field or on pre-set scales on the lens itself.

In 35 mm cameras, 28 mm focal lengths are considered to be almost telephotos when it comes to capturing the Moon's shadow effects across the sky. 24 mm is becoming a standard wide-angle in much ordinary photography, and these lenses perform well in about a 90° cover. 21 mm is even better.

Sequence photographs with 35 mm wide-angles can be done, although the image scale leaves a lot to be desired. Medium or larger format for this specialised topic is much better.

Fish-eyes are expensive, but justified for the unique pictures seen during totality. Their other main drawback is a relatively slow aperture, meaning longish exposure times, which is best left to autoexposure on the camera. Some medium format lenses of surprising quality can be obtained from Russian sources.

Many wide-angle photographs are ruined by sloping horizons, usually when there are vertical objects not spotted in the low light levels (see Fig. 10.2). These horizontal and vertical lines clash horribly, and become unacceptable if the camera is pointed up or down too much. A tripod is just as essential a piece of equipment as with a telephoto, and a spirit level attachment is of tremendous help when lining everything up during the excitement of totality.

Watch out for secondary imaging from the intense lighting at the ends of totality, and vignetting at the corners in sky-only shots. Make is no guarantee of absence of these faults, although they can add to the artistic style.

Telephotos

In normal parlance, anything much over 85 mm on a 35 mm camera tends to be known as a "telephoto" lens. At these modest focal lengths, very little can be done in the way of direct photos of the corona or other phenomena due to small image size and halation. Nevertheless, a 135 mm is considered the ideal focal length for portraiture and other candid shots, and enhanced landscape detail, and these lenses are often of superb quality.

Figure 10.2. Sloping horizons on wide-angled and fish-eye shots. India, October 1995. (Michael Maunder)

It is not until we get to what is often described as the extreme telephoto end that lenses become more suitable for eclipse photography. 300 mm is definitely a semi-wide-angle in this context and 500 mm = 5 mm (still less than 1/4 inch) on film, only mildly interesting. Image resolution is simply too poor, and halation still destroys much of that. The savage light intensity also shows up all optical faults in the way of ghost and other secondary images. If the reader finds a lens devoid of such faults, hang on to it, as it will be excellent for conventional photography. Baily's Beads and the Diamond Ring are possibly the ultimate tests of lens quality.

Glass vs Mirrors

The debate has been going on ever since Newton. There can be no doubt that glass lenses are easier to make, and now that autofocus and zooming are all the rage, the simpler telephotos for manual cameras can be bought very cheaply. Inspecting these takes little time and if unsatisfactory can be sold on for little loss, the point being that many have just the properties we need – simple construction and easy focus, and normally an end stop for infinity.

Nearly all these telephotos have a modest aperture, and when stopped down chromatic focus is not too bad. Even with focal multipliers on the end the performance can be acceptable, although the risk of secondary imaging increases alarmingly.

It is this problem of secondary imaging that needs most attention, because blooming and good internal blacking are not too critical in conventional photography. Distant terrestrial objects often have intervening mist, and a lack of contrast passes critical scrutiny! Modern telescopes and camera lenses increasingly apply aspherical optics, often with special glasses and better blooming and these should perform to specification.

Mirror optics, on the other hand, are quite a problem at an eclipse because they become unfocused extremely easily when the temperature changes. It is not just a question of the physical expansion and contractions of the tube and mounts altering collimation and focus, but the whole figure of the mirror(s) can go out of kilter. Night-time astronomers are only too familiar with the time taken to equilibrate.

It is a bad mistake to take too large an astronomical reflector, unless the image degradation and refocusing problems can be shown in practice to be kept within bounds. The high image scale will also be wasted and the full resolution lost in cutting-back to 35 mm format, unless only a small part of the limb is needed. Much better to take a small "spotting" reflecting telescope designed for cameras as these will have a much slicker focus arrangement which can be pre-calibrated.

To summarise: simple optics are more important than a grand scale to impress your friends. Glass optics are robust and give excellent contrast and resolution even on afocal projection; mirrors (without glass lenses for close focus) give guaranteed chromatic focus at all colours, and can be made to work at faster f ratios because of this. These two properties are not mutually exclusive.

Use glass lenses for high resolution and mirrors for the fainter extended objects such as the corona.

Optimum Focal Length

In commenting on mirrors above and in Chapter 8, sheer image size is not the only criterion – quality under field conditions has to come into it. The longer the focal length chosen, the more care has to be taken to check the system as a whole for camera shake and stability. Also, will it fit comfortably into your luggage?

It is worth repeating that you should choose only the longest focal length conveniently working within your chosen film or video format, i.e. the image size must be large enough to be useful but not so large that it is difficult to keep within the field of view. In practical terms this usually means a focal length around a metre (1000 mm) for 35 mm, which gives an image size close to 10 mm. This gives about half a solar diameter clearance either side across the width of the film, usually enough to compensate for any slack in the mounting when brought to the field centre and released, and still show enough corona to be interesting.

An equatorial motor drive makes life so much easier that one ought to be included in your list of essentials if transport problems can be solved. Without one, a focal length of ~1 metre then comes into its own, as it is relatively easy to arrange the mounting so that the solar image drifts down the long length of a 35 mm frame or video screen. As a first approximation, this image drift takes two to three minutes whilst still keeping all the solar image in the field of view. For many of the shorter events this is more than enough time to get all the pictures you need between second and third contact. It is certainly long enough for routine solar observation.

The time restraints are considerably eased if a larger format camera can be used. It takes a good five minutes for a metre focal length image to drift far out from the centre of a 6 × 6 cm film frame, once started close to an edge. Appendix E lists some other transit times.

Zoom Lenses

Zoom lenses are fine for their original purpose, which is to cut down the number carried for ordinary photography. By their very nature they have to be a compromise. Avoid them wherever possible for any main experiments, even with modern optics, and always select instead the best prime lens you can afford. Quite often a simple glass lens will outperform a zoom for reasons not relevant here. Suffice to say that the risk of ghosts and other secondary images and lowered contrast is too high. See Fig. 10.5(a).

Tele-extenders

Do not add tele-extenders until the particular combination has been tested fully. There is a severe light loss and image degradation, usually with a much poorer contrast. By far the more serious effect is the worsening of camera shake as the lens becomes unbalanced.

A 2 × extender reduces the light by two stops, and to cope with the image degradation some further stopping is necessary. Often a longer focal length lens will perform better, and as noted above, these can be cheap.

Filters

All telephotos, telescopes or other optical aids need some form of solar filter and even if this is lightweight Mylar, it can unbalance the lens. Lining up on the Sun with a dense filter is surprisingly difficult, particularly in the haste often necessary on a crowded eclipse site. Spend some time and effort designing a finder. In its simplest form it could be a back-up camera which is used as a counter-weight at the same time (see Fig. 10.3).

Check that alignment is consistent whatever device is chosen within the intended accuracy and that it does not snag vital parts at the anticipated elevation. Pay particular attention to balance and drive irregularities.

Focusing

There is one final check to make on the optics, particularly if the focal length is long. More eclipse pictures are ruined by poor focusing than any other factor except camera shake. Too many published pictures show this problem, and there are two simple options which come down to the same thing in the end – work on it back home, not on the day.

Pragmatism

Where there is difficulty using a camera viewfinder without some optical aid like glasses, and/or a correct-

Figure 10.3.
Counterbalancing
cameras to reduce
camera shake.
(Michael Maunder)

ing lens cannot be fitted, make test shots a little out of focus either way and see if it improves the image sharpness. If it does, make a note on the lens where *actual* focus occurs. Then check if there is a change each time you reassemble the set-up. It will use up some film, but better now, and not when the exposures are for real.

The test object does not have to be the Sun; the Moon is a lot safer. Colour rendition and focus can be checked with a far distant street light, where grossly over-exposing on a distant bright light source will also pick up any tendency to ghost imaging.

Science

Mirror "lenses" should focus all colours, but with the tendency to add glass elements for close focus etc. it is always a good idea to run these through a colour check as well.

The basic problem is that at totality, the corona has a pretty blue colour, some green and yellow is also evident, and the prominences are almost into the infrared, a range which demands a lot of any optical system. Most camera lenses will have an infrared mark and explain how to use it. If this mark is well away from the zero, we have a problem even on stopping down and an alternative lens might be the only answer. The lens will need to be stopped down at least as far as the depth of field indicator.

The classic way of testing colour correction is to focus visually on a multi-coloured object and place filters in between. Any need to refocus with one of the trichromatic Wratten filters identifies a problem straight away. The filters to use are 25 (red), 58 (green) and 47B (blue).

The modern way is to illuminate a high resolution bar chart with different colour LED light, which is almost monochromatic. Again, shifts in focus can be picked up quickly and used as a simple sorting test.

Camera Shake

Since camera shake is by far the most important single cause of degraded images in solar photography with conventional cameras, it is worth reiterating some of the points made earlier. It can also happen with video, and for the same reasons. Heavy lenses and attachments stuck on the front of the camera upset its natural balance and the slightest vibration in the mechanics start a nodding and rocking vibration. Even with perfect mechanics, wind buffeting can, and usually does, play havoc. See Fig. 10.5(b).

A very sturdy tripod becomes the most essential ingredient in the whole system. It cannot be emphasised enough that money spent on a good tripod or support is often a much better investment than a fancy lens. Effectiveness is the only important criterion.

Tripod sturdiness must be checked for the longest focal length lens, and is best done by test exposures. But before that expense, tap the lens. Does the image take longer than 1/10 s to settle down? If so, the structure is unsuitably flimsy. Try the effect of adding weights or bracing struts. Also try the effect of a hair-drier at close range to simulate strong wind buffeting.

If the tripod passes the simple tap test, it is still possible to get camera shake because of the instability of the optics stuck on the front of the camera, when the only mounting point is the standard camera bush. Whenever camera shake is encountered, always check the natural point of balance of the system. If this is not where the camera is mounted on the tripod, you can forget about sharp images, unless you are very lucky, because of the natural vibrations about the incorrect point of balance. Somehow a means of mounting the camera at its natural centre of balance has to be used, even if this means using a different optical system with a correct tripod bush. Telelenses without their own tripod bush are unacceptable.

Exposures in the most useful range of 1/125 to 1 s give most bother. Try some test exposures in this range. Do not put real film in the camera, just a dummy film in to get the balance right. If the front point of the lens jumps about after exposure, camera shake is a real problem.

Having made sure the equipment will sit on the mount comfortably and without vibration, the final check is to see if it lines up with the Sun at the anticipated altitude and azimuth without snagging something, or worse. This final point should be blindingly obvious, but far too many people miss out this killer. All too often the camera equipment performs beautifully when pointed straight ahead, but then bounces around like a ship in a gale at an extreme elevation. The unit has to perform as a unit with all the counterweights in position. The weak point is more often than not the pan and tilt or universal head which are not made with astronomical elevations in mind.

Provided all the basics of tripod sturdiness and balancing are attended to, the effects of camera shake should be minimised to a surprising extent. However it still pays to relegate any camera with a marked "jump" to wide-angle and standard lens topics. Secondly, any camera with a motor drive can induce vibrations for a long time, and it might pay to consider them as a wind-on aid only, rather than for rapid sequences.

By far the best type of camera to use is one with a mirror lock, as the effects are quite dramatic. Time spent devising an alternative viewfinder, such as a collimated pair of binoculars, then pays off. The loss of the direct vision screen is more than offset by the better resolution and scale of the direct view, and these used to trigger exposures that fraction of a second earlier, a point which can make all the difference in the final image quality.

Suggested Observations

Many people just turn up on site and play it by ear, but for most of us a little more organisation pays dividends.

Partial Phases

Nothing else really matters except the precious moments of totality, and provided the travel arrangements don't mean leaving immediately afterwards, consider doing this work then and kid everyone it's the same in reverse.

The weird lighting becomes more obvious, shadows are still there but much more diffuse and they take on a distinctive greyish hue. In the last few minutes before totality, the sky itself near the Sun takes on a unique steely blue tinge, with reddish to orange shading towards the horizon. If you have a spare camera or video, it is well worth thinking in terms of just simply recording the darkening sky. The modern automatic exposure systems cope very well, and the pictures will give some impression to others later, and bring back powerful memories for those at the event. It is also a sensible back-up in case the weather turns nasty for totality.

Manually-operated cameras are even more useful, as they record the actual lighting intensity much better. The light levels fall off roughly in proportion to the area of the Sun; thus a 50% phase meters in at around one stop down and so on until the last few per cent when the metering drops off the scale to night-time levels.

Shadow Bands

If you are really lucky, with crystal clear skies, shadow bands might occur. More observations are needed with timings to finally clear up what they are and how they are formed. They are seen as a diffuse pattern of light and darker shadings moving rapidly over any clean flat surface such as whitewashed walls, sheets or even large sheets of papers. Shadow bands only appear a minute or so either side of totality and many experienced photographers describe capturing the effect as the ultimate test of skill and luck (see Fig. 10.4).

Baily's Beads and the Diamond Ring Effect

After totality, the Beads are always less dramatic but as the lighting increases rapidly one or more of the Beads develops into a full Diamond Ring. A small amount of over-exposure will emphasise the Ring, but do not overdo it.

The Diamond Ring effect has been observed just before totality, but is always more spectacular just afterwards. Photography confirms that impression, so it is not just a mental optical illusion combined with some eye dark adaptation. The effect is so startling, some say beautiful, that many observers travel to see just this second or two of Nature at its most exquisite. The Diamond Ring effect signals the end of totality, and that full eye protection is needed again until the next total eclipse.

Figure 10.4. Liverpool Astronomical Society trying to photograph the elusive shadow bands in Kenya, 1980.

The Chromosphere

Just before Baily's Beads appear an amazing drop in light level makes the ground lighting much redder in an almost smooth transition from the reddening limb colour to the pink of the chromosphere proper.

Seen from Earth the thickness is but a small fraction of the diameter, and in long eclipses may be seen only at the leading and trailing edge of the Moon's disk. Some observers sacrifice maximum duration in order to capture more striking pictures with the chromosphere more in evidence at one of the other limbs.

When the Moon's diameter closely matches that of the Sun, the chromosphere can be seen right the way round and is a sight in itself. It is sometimes seen during annular eclipses. On top of the chromosphere, and using the same sort of exposures are the prominences.

The Prominences

Catching the prominences places a premium on optical quality and above all camera shake and accurate focus (see Fig. 10.5).

The colour is a strong pink to red, and being monochromatic, must be a disappointment to about a third of the male population with colour blindness. Many monochrome films are also blind at these wavelengths.

The shape and structure vary enormously, as does the number seen. There is always a greater number at sunspot maximum and when spots are on the limb. Some eclipse chasers spend much time plotting sunspot groups weeks before the eclipse in the hope of catching prominences at the right time and place on the disk on the day. Others let Nature take its course for a pleasant surprise.

The Corona

With the possible exception of the Diamond Ring, all observers agree that there is nothing to compare with the corona in full display. Even hazy/cirrus clouds cannot detract from its full wonder.

The colour is a pearly pink/grey to those with perfect colour vision. The tracery, filamental fine structure and overall shape vary according to the sunspot cycle,

a

b

Figure 10.5.
Examples of bad photographs from Michael's "learning curve"! **a** Typical result of flare from poor contrast optics, or inadequate blooming. This example results from poor blackening in the tube of the mirror optics. **b** A good picture ruined by camera shake. Prominences and inner corona are "smeared" onto the lunar disk, and show no detail.

being more circular at maximum, and with sharply delineated wings at minimum with plumes and spikes at angles from the centre. The naked eye can usually see out to more than a solar diameter (angular $1\frac{1}{2}$–2°), particularly if dark adapted. Any form of optical aid is always recommended, and with a modest pair of binoculars the corona should be visible out to several solar diameters. It is the only way to fully appreciate the delicate structure and stunning beauty.

Any attempt to catch the full corona has to make allowance for the considerable area it covers – often five to six solar diameters out or more than 5° across. Few cameras and lenses can cover that extent at an extreme focal length, and even video technology has difficulty coping with the full intensity range from the equivalent of daylight to the night sky. There are special techniques for grading filters, afocal stops and similar ways of overcoming the problem, and all require meticulous planning to get the conditions right.

Other Phenomena

If you can tear your eyes away from the corona, spare a moment or two to look about. The Moon's shadow often appears quite clearly; "set into the sky" is one way of describing it. Sometimes it seems to be like a doom-laden tunnel rushing towards you across the landscape. That and the strange reddish horizon and edges to the shadow make this a scene never to be forgotten. It is best seen across a clear landscape, as on the snow fields of Montana in 1979, and in Mongolia 1997, a Saros later.

During totality the brighter stars and planets will appear, often a good time before. A wide-angle photograph is always something to attempt. Charts of their positions are always available in the major magazines to plan ahead.

Even more important to some people is the sky darkening at a time not normally available. Comet hunters frequently spend much time searching in the hope of repeating the discovery of 1948. Eclipse chasers in Mongolia were clouded out in March 1997, but China was luckier, and had excellent views of comet Hale–Bopp.

Choice of Recording Medium

Video and camcorder auto-exposure functions will cope with virtually any situation. Buy only the best quality tapes when recording the subtle coronal detail, and take plenty of them for everything else – don't try to economise or compromise.

It is not so simple with still photography, because of the huge range. Stick with whatever is a favourite make or type for normal use, as the film's properties will be known.

The general rules are very simple:

1. Slow films give better grain and colours. Use the slowest film possible if the optics and camera shake problems have been overcome. Go up to 100 ISO for prominences and the inner corona if there is any doubt on this score, as the slightly worse granularity and halation will mask the problem and allow shorter exposures. Only go to 400 ISO or higher if you are in real trouble.

2. Fast films record faint detail where quality is of secondary importance. Use 400 ISO or faster for the outer corona.

3. Print film is always more forgiving than slide, and where exposure time is at a premium, go for print every time to save bracketing.

4. If the situation allows, always double-up equipment or duplicate it as a back-up. If all works, bracketing is not necessary as a different film type or speed can be used instead. Double cable releases, electronic alternatives and pneumatic devices all work well.

Finally

Take at least twice as much as you think you will need, and then some more. Set aside all the film you think necessary for the eclipse itself and store separately.

Test Objects

Whilst test exposures of the Sun are important for the partial phases, they are not the best check on tripod and camera stability. Exposure times are too short even with a decent filter.

As mentioned earlier, by far the best test object is the Moon, which is the right size and brightness range, and the apparent motion is also close enough to check guiding or motor drive errors.

Full moon is about the brightness of the inner corona. A few exposures will be a useful double check that the f-ratio of the optical system is what is claimed. Full moon, is, by definition, a day (sun) lit object. The crescent moon is the best test object for focus and camera shake checks. Allow two stops extra exposure to pick up crater detail, and do duplicate exposures during periods of good seeing. It will probably take some lunations to iron out all the bugs.

Do not always expect that the image in the viewfinder is absolutely correct focus. Follow the pragmatic approach above, and make test shots a little out of focus either way and see if it improves the image sharpness. If it does, make a note on the lens where *actual* focus occurs. Does this change each time you reassemble the set-up?

Exposure Planning

Video needs the same discipline as still or ciné. Plan a sequence of exposures, keep it simple, and write it down. Work through this plan in a dummy run (no film in the camera), with no more room lighting than the full moon. Allow at least 50% of the time to just stand and stare. Use that 50% now to note the time taken and write it against each step. An auto-wind camera with a timer does make life easier these days.

Keep at it until you have a plan capable of being carried out three-quarters of the totality time, a sensible safety margin for more naked-eye staring.

Simple plans can be committed to memory, but the panic of last minute hitches makes this a dangerous policy. It is much better to take along a simple "Walkman"-type player. Use your script and a stop watch and actually record the plan, in real time, and work through the sequence using your own voice prompts. Provided you are well away from others on the site, another recorder is a great idea to take down your verbal record of what you actually do on the day. However, do bear in mind that other workers might not like your commentary. The advantage of your taped prompt is that it is directly into the ear and blots out other voices.

The Plan

Knowing what to do at an eclipse is always the most difficult thing to decide. There are simply too many variables of equipment and weather to be dogmatic, and this is merely a suggestion which has stood the test of time.

1. Always take plenty of pictures of the site as you arrive and set up. That is why you need lots of fun cameras. Most eclipse viewers pack up immediately after totality and are never seen again!

2. If you only have one main camera, change film at the first opportunity to the right one, check the settings, then leave well alone.

3. Set up the equipment. Check that all is in order and do the necessary if there are problems. Use your audio prompt.

4. Take the absolute minimum of pictures during the partial phases, remembering that sunspot occultations are fascinating and you might want to change your mind. You will know if this aspect is a possibility long before first contact.

5. Take lots of pictures with your back-up and fun cameras of the pinhole effect and any other phenomena, such as birds going to roost or flowers closing.

6. One minute before second contact, double check that the camera is correctly set for Baily's Beads.

7. Judge the Baily's Bead stage, and take lots of pictures.

8. Take plenty of pictures during totality, not forgetting to look around and at it as well. Annularities close to total might show some corona. This was certainly true in 1984, and there are records of prominences being seen well before and after totality.

9. Take more pictures of the second set of Baily's Beads at third contact, and with luck these should come out as the Diamond Ring if the camera settings cannot be changed.

10. Re-look at the partial eclipse, with optical filters, of course, and the general scenery.

11. Repeat Instruction 4.

12. Relax or finish the film on pinhole effects or the scenery, wildlife or whatever you choose. Many of the celebrations make memorable pictures, as in Fig. 9.1.

13. Double check dismantling and packing procedures, autocueing if necessary.

14. Double check that films are correctly labelled and stowed away for the journey home.

This suggested plan is amongst the simplest, but, even then, has fourteen important steps to carry out. The need for practice beforehand should be abundantly obvious.

Chapter 11

The 1999 Eclipse

With the possible exception of comets, the total eclipse on 11 August 1999 will be the most important astronomical event for a long time in the UK and much of mainland Europe. For the Channel Islands, not in the UK but part of the wider British Isles, the eclipse is even more momentous.

The only thought to bear in mind at any eclipse is painfully obvious – see it! Anything that makes that task easier has to be a plus point, and knowing when to be ready is part of the skill. However, the biggest mistake of all is to become obsessed with timing too accurately, as only dedicated researchers have to work that way, and they will have their work programme sorted out long before the event. Modern computer programs can time all the phenomena to fractions of a second, but this level of precision is somewhat of an overkill for the average person only interested in glimpsing the more interesting bits before returning to something else, and many astronomers fall into this category too. No, what we want is something which helps us to be *ready* for the events.

Experience has shown that it is much better to be ready a few seconds before the major events, rather than to home in on the precise time. The reason is very simple; slight errors in the programme or geography can throw the timings out enough to unnerve the inexperienced. There is also nothing worse than making a mistake in setting the clock, and then missing the critical event by a second or so.

What we need is a pocket aid of the basic times, and the tables that follow set out to do that. Times for first, second and fourth contact are listed a minimum of 5 to 10 seconds early, and quoted to the nearest start or half of the relevant minute as this is always the easiest thing to see or set in advance. Just make sure that the clock or timer is set as accurately as possible beforehand from the Rugby radio, Teletext etc. In some cases this might mean being ready for over half a minute when the event falls less than 10 seconds after the next relevant start or half minute, which is no bad thing, as the all the phenomena associated with second and fourth contacts can be seen to be changing rapidly and so lead you more gently into what amounts to a useful secondary warning.

In the unfortunate event of cloudy weather, part of the folklore of eclipse chasing is the (apparently) magical clearance which takes all unawares. All the more reason then for knowing when to be ready in advance as those extra seconds can make all the difference in being prepared, psychologically, as well as practically.

The exception to this guideline is the timing for central totality, which is quoted to the nearest second or so as this is the only critical thing to know. It has to be that accurate, as it often pays to set a timer to tip off the central point when checking an observing programme. If way behind schedule, for whatever reason, give into the luxury of a panic, and go straight to naked-eye observations. As mentioned many times before, build that fall-back position of at least 50% into the programme. Experienced observers will confirm that they never regret doing this, but only that they never saw totality because they were trying to beat the odds when behind schedule.

Mid-point timing becomes important to anyone making a sequence of pictures, and can be used to calculate starting and ending times to get the whole image on a frame. Auto-exposing cameras make life so much easier, and the authors have used this feature to set apparatus the night before. As long as the equipment is in position at an early stage, the electronics can be allowed to work without further worry.

Third contact time will also seem to be a few seconds late, for a very good reason. If you have not repositioned the dense solar filters by this stage, you will be in very real trouble indeed. Re-read the warnings about the dangers and take every precaution to act responsibly. **Subtract 10 seconds** to expect the Diamond Ring, or use totality duration to work from second contact as it occurs on the day.

GMT is used throughout. Adjust timings according to the appropriate summer time in each country.

It is obviously not possible to give information for every place within totality in a book of this type and quite impossible for the huge number of places experiencing a large partial phase. The selection is based on major sites as the authors see it, such as their home towns, or where the British Astronomical Societies are settling down to see it. Figure 11.1 shows the track of totality over Britain.

The Royal Astronomical Society will be holding their week-long National Astronomy Meeting in Guernsey, and travelling to Alderney, MM's home, to view the eclipse. The British Astronomical Association will be based in Truro and, depending on travel restrictions and weather forecasts, will use Goonhilly Down or Falmouth as alternative sites where the timings are very similar. Many other societies and the media will be moving down to Cornwall and the Scillies for the event.

The weather prospects in the Peninsula can be very difficult to predict. August is probably best described as the UK "monsoon season", with often as much rain as any other time of year, much of it falling on only a few days. The problem is one of thunderstorms, enhanced by hot humid air being forced to rise over cliffs and other high ground in Cornwall. These clouds

Figure 11.1. The track of totality over Cornwall.

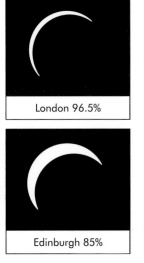

London 96.5%

Edinburgh 85%

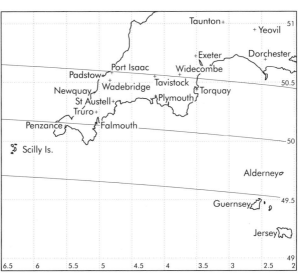

Table 11.1. Conditions for totality

GMT	First contact	Second contact	Mid Totality	Third contact	Fourth contact	Duration min.sec.	Altitude degrees
Capitals Along the Track							
Scilly Isles St Mary's	08-55-5	10-09	10-10-30	10-11-30	11-30	1-42	45
Cornwall Truro	08-57	10-11	10-12-24	10-13-30	11-32	2-01	46
Alderney St Anne	08-59	10-15	10-16-08	10-17-10	11-37	1-47	48
Luxembourg	09-09	10-28	10-28-54	10-29-40	11-50-5	1-14	53
Strasbourg	09-10-5	10-30-5	10-31-30	10-32-20	11-54	1-22	54
Bucharest	09-41	11-05	11-06-54	11-08-15	12-28-5	2-22	59
Karachi	11-17-5	12-25-5	12-26-24	12-27-10	13-27	1-13	22
Famous Places Along the Track							
UK							
Land's End	08-56	10-10	10-11-20	10-12-30	11-31	2-03	46
Helston	08-56-5	10-10-5	10-12-00	10-13-10	11-31-5	2-06	46
Lizard	08-56-5	10-10-5	10-12-01	10-13-10	11-32	1-59	46
Plymouth	08-58	10-12-5	10-13-41	10-14-40	11-33-5	1-42	46
Dartmouth	08-58-5	10-13	10-14-28	10-15-25	11-34-5	1-44	47
Torquay	08-58-5	10-13-5	10-14-39	10-15-20	11-34-5	1-12	47
Europe							
Cherbourg	09-00	10-16	10-17-06	10-18-05	11-38	1-40	49
Reims	09-06	10-24	10-25-30	10-26-35	11-47-5	1-59	52
Munich	09-16	10-36-5	10-38-12	10-39-25	12-01	2-07	56
Salzburg	09-18	10-39-5	10-40-48	10-41-55	12-03-5	2-02	57

For First Contact, time to be ready for "first bite"
For Maximum Phase, time to be ready for it
For Fourth Contact, time to be ready for "last chip"

begin to build up during the morning as the ground warms up and sucks in air from the sea, just the timing of most concern for totality.

Many valiant efforts have been made to survey the records, but these examinations often predict poor prospects in much of the Peninsula on the eleventh in any ten-year span to date. The site to seek is towards the sea, preferably on a promontory, and fortunately the Cornish coast has plenty of these. The real sticking point is going to be travel prospects, with little or no chance to move if the weather changes, as early indications are that the authorities will place a total ban on traffic. This will be a sensible arrangement as essential services must have freedom of movement.

The best promontory of all is an island, and the Scillies could well offer the best prospects of all.

Table 11.2. Some capitals in partial eclipse in 1999

GMT	First contact	Maximum phase	Fourth contact	Altitude degrees	Obscuration percentage
		British Isles			
Belfast	09-01	10-13·5	11-30	43	89
St Peter Port	08-58·5	10-15	11-36	48	99.8
Cardiff	09-00	10-15	11-35	46	97
Douglas	09-02	10-15	11-32	44	90
St Helier	08-59	10-15·5	11-37	49	99.1
Newport (I of W)	09-01	10-17	11-38	48	99.1
Inverness (Highlands)	09-06	10-17	11-31	42	81
Edinburgh	09-05	10-17·5	11-33	43	85
		Europe			
Dublin	08-59	11-12	11-30	43	93
Paris	09-03·5	10-22	11-44·5	52	99.4
Zurich	09-11	10-32·5	11-56	56	97
Rome	09-17	10-42	12-08	63	84
Zagreb	09-22	10-47	12-11	59	97
Vienna	09-26·5	10-54	12-20	68	68
Budapest	09-28	10-51	12-13·5	58	99.4
Belgrade	09-30·5	10-56	12-19·5	60	98
Sofia	09-36·5	11-03	12-27	62	94
Athens	09-41	11-10	12-34	66	82

For First Contact, time to be ready for "first bite"
For Maximum Phase, time to be ready for it
For Fourth Contact, time to be ready for "last chip"

Which leaves Alderney. 1996 was the first year in a decade when the weather was not bright and clear at the critical time, and that did allow enough of a glimpse to photograph prominences through the scudding clouds. The important feature about Alderney is its dryness compared with the other Channel Islands due to its shape, often having dryness or clear skies when Guernsey and Jersey get rain. It is long and thin and pointing into the prevailing SW wind, but more important in our context, nowhere is over 280 ft (180 metres) so that air is not so pushed up to form storm clouds. It makes for almost ideal astronomical seeing with laminar flow conditions.

The two lunar and the solar eclipse of 1996 were seen from Alderney in good conditions.

Special mention needs to be made of those places just outside totality, because this is a relatively long eclipse with good lunar cover. Jersey and particularly Guernsey are almost in totality and experience has shown that with good clear conditions, and a bit of

Table 11.3. The UK partial eclipse conditions in 1999

GMT	First contact	Maximum phase	Fourth contact	Altitude degrees	Obscuration percentage
The Really Important Places					
Selsey	09-01-5	10-18	11-38	48	99.1
Basingstoke	09-02	10-18	11-38	48	98
Clapham	09-03	10-19	11-39	48	97
Greenwich	09-03	10-19	11-39	48	97
Tunbridge Wells	09-03	10-20	11-40	49	98
Watford	09-03	10-19	11-39	48	96
Famous Places					
Lundy	08-58	10-13	11-32	46	98
Widecombe	08-58	10-14	11-34	47	99.99
Bath	09-00-5	10-16	11-36	47	98
Blackpool	09-03	10-16-5	11-34	45	91
Cowes	09-01	10-17	11-38	48	99.3
Stonehenge	09-01	10-17	11-37	47	98
Birmingham	09-02	10-17	11-36	46	94
Stratford-on-Avon	09-02	10-17-5	11-37	47	95
Newbury	09-02	10-18	11-38	48	97
Brighton	09-02	10-19	11-40	49	99
Luton	09-03	10-19	11-39	48	96
Scarborough	09-06	10-20	11-38	46	89
Norwich	09-06	10-22	11-41	48	93

For First Contact, time to be ready for "first bite"
For Maximum Phase, time to be ready for it
For Fourth Contact, time to be ready for "last chip"

luck, it might, just, be possible to pick up some of the prominences if particularly large or bright, and perhaps some inner corona. The limb profile does not seem to be all that good for Baily's Beads, but you never know.

Whilst the difference between totality and 99% is chalk and cheese, those who cannot travel to totality can rest assured that the views will still be worthwhile, and possibly of some scientific interest.

Chapter 12
Eclipse Mishaps and Oddities

Eclipse stories are many and varied. Some of them illustrate what is often referred to as Spode's Law: If things *can* go wrong, they *do*! Of course, the main hazard is cloudiness. There have been countless eclipse expeditions which have been wrecked by overcast skies, and there is absolutely nothing which can be done about it.

All precautions are taken, but the weather is notoriously unreliable in most parts of the world. This was emphasised at the eclipse of July 1991. Two of the most favoured sites were Mexico and Hawaii. Totality in Mexico was longer, but cloud was a distinct possibility, and so most of the main parties went to Hawaii, which was expected to be crystal-clear. In the event Hawaii was largely clouded out, while conditions in Mexico were perfect.

The first American eclipse expedition was not a success. The eclipse fell on 21 October 1780, and a party led by Samuel Williams went from Cambridge (Massachusetts) to Penobscot in Maine, where conditions were expected to be favourable. Clouds were absent, but unfortunately there was a slight miscalculation, and the astronomers stationed themselves just outside the zone of totality. They watched as the Moon's shadow crept across the Sun – and continued to watch as the Moon drew back once more. Gloomily, the expedition packed up its equipment and went home. (There was a similar fiasco in September 1941, at the height of the war. A French expedition managed to get as far as the Pacific, but ended up on the wrong island – again outside the belt of totality.)

The least successful eclipse manoeuvre stands to the credit of General William Harrison, later President of the United States, when he was Governor of Indiana Territory. He was having a difficult time with the Shawnee Indians, who had absolutely no wish to see white men in their backyard. The most influential Shawnee prophet was named Tenskwatawa. Harrison decided to ridicule him by asking if he could "cause the Sun to stand still and the Moon to alter its course". Alas, the prophet knew more about astronomy than the General, and he was well aware that on 16 July 1806 there would be a solar eclipse. He therefore proclaimed that he would demonstrate his power by blotting out the Sun, and at the appointed time a large crowd gathered to mark the occasion. When the eclipse began, the effect on the American troops can well be imagined, and the General was taken aback. In 1811 he did indeed destroy the Shawnee power at the Battle of Tippancoe, but by then it was rather too late to rescue his scientific reputation.

By the mid-nineteenth century photography was starting to come to the fore, and at the eclipse of 1851 an astronomer named Berkowski took the first picture to show the corona and the prominences. Others were also in the party, and it is on record that one enthusiast forgot to load his camera. This sort of thing has happened many times since, and there are other traps too. Both the present writers took part in an eclipse expedition to the African coast in 1973. Conditions were excellent, and this was a particularly long eclipse; it lasted for seven minutes, which is only slightly short of the theoretical records. One member of the party took forty exposures of the corona under ideal circumstances; only after the end of totality did he find that the lens-cap of his camera had been firmly in position throughout. Not unnaturally, he was much displeased.

The eclipse of 22 December 1870 was expected to be of special interest, and one man who was determined to make the most of it was Jules Janssen, a leading French astronomer who made notable contributions to solar physics (it was he who, simultaneously with Norman Lockyer, first realised that prominences could be studied spectroscopically at any time, without waiting for an eclipse). Unfortunately for Janssen, Paris was being beseiged by Germans at that time, effectively isolating him. But, undaunted, he equipped himself with a balloon, and made a spectacular escape from the city. He landed safely, and made his way to the chosen observing site at Oran, where he was greeted with totally overcast skies.

This may have been frustrating, but in 1886 a Russian astronomer named Mendeleef had an even more bizarre experience. Fearing clouds, he emulated Janssen and decided upon a balloon ascent, and engaged the services of a professional aeronaut. All seemed to be well, but at the vital moment the balloon suddenly rose upward leaving the aeronaut behind – and the bewildered professor found himself alone, sailing gaily upward to a height of over 11,000 feet. Fortunately he knew enough to control his flight, and landed unhurt, but though he had a magnificent view of totality he was rather too preoccupied to make any scientific observations.

Only during totality can stars be clearly seen in the daytime, and this fact has been used on several occasions. The first seems to have been in 1878, and the story is certainly worth retelling.

It really goes back to 1846, with the discovery of the planet Neptune. William Herschel has discovered Uranus in 1781, during a routine "survey of the heavens" with his home-made telescope; it soon became clear that Uranus was misbehaving, because it persistently wandered away from its predicted path, and astronomers were very anxious to find out why. Two mathematicians, John Couch Adams in England and Urbain Le Verrier in France, independently concluded that the cause must be gravitational pull of a more remote planet. They worked out where the troublemaker ought to be and, in 1846, Johann Galle and Heinrich D'Arrest, at Berlin, found Neptune almost exactly in the predicted position.[*]

Le Verrier then turned his attention to Mercury, which also appeared to be straying. Le Verrier concluded that there must be yet another unknown planet, this time moving well within the orbit of Mercury, and he made calculations of the same sort as he had done for Neptune. The real problem was that even Mercury, with its mean distance of 36 000 000 miles from the Sun, is a somewhat elusive object, and Le Verrier's planet would certainly be so close-in that it would be virtually impossible to observe unless it could be caught in transit across the Sun's disk or else located during the darkness of a total eclipse.

---

[*] For the full story, see *The Planet Neptune*, by Patrick Moore. New edition: Praxis Press, Chichester (1996).

In 1859 a French amateur named Lescarbault claimed that on 26 March of that year he had actually seen the planet in transit, and Le Verrier made haste to go and see him at his home at Orgères, in the Department of Eure-et-Loire. It must have been a curious interview. Le Verrier had the reputation of being one of the rudest men who has ever lived; Lescarbault, who doubled as a local doctor and carpenter, had a small telescope, recording his observations on planks of wood and planing them off when he had no further use for them. Yet Le Verrier came away fully convinced, and he even gave the planet a name: Vulcan, in honour of the black-smith of the gods. According to his calculated orbit, Vulcan moved round the Sun at a mean distance of 13,082,000 miles in a period of 19 days 17 hours.

True, another French astronomer – E. Liais – had been observing the Sun at the same time, and had seen nothing at all; but this did not dampen Le Verrier's enthusiasm, and his faith in Vulcan remained unshaken right up to the time of his death in 1877. Meanwhile, the eclipse of 29 July 1878 seemed to provide a good opportunity for Vulcan-hunters, and two of those who took up the challenge were Professor J.C. Watson, a well-known American astronomer, and Lewis Swift, who had earned a great reputation for dis-covering comets. (Comet Swift–Tuttle, parent of the Perseid meteors, was one of his finds.)

The skies were clear, and both Watson and Swift carried out careful surveys of the region of the hidden Sun. Each recorded unidentified objects, but their observations were discordant, and if they could be regarded as reliable there was no escape from the con-clusion that they had found not one Vulcan, but several! Alas, there seems no doubt that they recorded nothing but faint stars. Vulcan has never been seen at any subsequent eclipse, and much later the irregulari-ties in the movements of Mercury were satisfactorily explained by Einstein's theory of relativity. Today there is full agreement that Vulcan does not exist; there is no large planet moving within the orbit of Mercury, and only comets and occasional small asteroids venture into this torrid region.

However, two comets have been found during total-ity. The first was in 1882, when pictures taken by an eclipse expedition in Egypt showed a bright comet close to the Sun; it was often referred to as Tewfik's Comet, in honour of the then Egyptian ruler. It had never been seen before, and it was never seen again, so that we know practically nothing about it. At the eclipse of 1 November 1948 another unexpected comet was found, and was recovered later; apparently the first man to observe it after totality was an aircraft pilot, Captain Frank McGann, who was flying over Jamaica. Eventually the comet reached the second magnitude, and developed an appreciable tail. It was followed telescopically until the following April; it will be back once more in around 95,000 years.

In 1919 there was a different sort of experiment. Einstein's theory had predicted that stars near the Sun should be slightly displaced – and of course this could be checked directly only when the Sun was hidden. Accordingly, an expedition led by Professor (afterward Sir) Arthur Eddington set out to make the relevant measurements – and the results were in full agreement with prediction; this was in fact the first purely observational confirmation of Einstein's theory. (It is on record that Eddington's assistant asked the Astronomer Royal what would happen if the results failed to agree with Einstein. "In that case," said the Astronomer Royal solemnly, "Eddington will go mad, and you will have to come back alone.")

Come now to near-modern times. There was a total eclipse on 15 February 1961, and BBC Television made arrangements to show totality three times as the shadow passed across the Earth's surface – first in France, then in Italy, and finally in what was then Yugoslavia. On of us (PM) was deeply involved, and was dispatched to the top of the Yugoslav mountain Jastrebač in the hope of being above any awkward clouds. The following is an extract from the observing log written at the time:

> Undoubtedly Spode's Law was in full operation, apart from the fact that the sky was clear in France and Italy and partially clear in Jugoslavia. From France, the pictures were well received, and were shown "live", but unfortunately the commentary, by Dr Butler, was lost; he could be seen talking, and his voice actually reached the BBC studio, but the floor manager omitted to plug in the microphone. In Italy, the commentator – Colin Ronan – has just described the "deadly hush" as totality approached, when the waiting crowd gave a roar of applause.
>
> On Mount Jastrebač, our equipment had been hauled up the slope by mountain oxen. Unbeknown to me, the Jugoslav director was a man with ideas of his own, and communication was not too easy; in the end I talked French to a German astronomer, who relayed it

to the director in Serbo-Croat. The director knew that animals tend to go to sleep as soon as darkness falls, so he arranged for cameras to be turned on to the oxen at the moment of totality. Just to make sure that everyone could see them properly, he floodlit them...

One final oddity concerns the eclipse of 1980, visible from India. An important cricket match was in progress – a full Test between India and England – and by mutual agreement the date of the eclipse was made a "rest day", because the Indian Board of Control did not want to shoulder the responsibility of a crowd of 50,000 people damaging their eyes by looking straight at the Sun when the eclipse began.

No doubt there will be many more eclipse stories during the coming years. But all in all, most mishaps are due to sheer lack of preparation. When setting out to photograph a total eclipse, on no account fail to make sure that your camera is loaded and in full working order.

Chapter 13

Eclipses of the Future

Quite a number of solar eclipses will occur during the first two decades of the next century. The following three tables list all solar eclipses occurring between 1998 and 2020, more than a Saros period.

- Table 13.1: Total eclipses. Times are given, together with the position on Earth where maximum eclipse will be seen (useful if, for example, you are planning an eclipse expedition in the Pacific Ocean!). The "ratio" indicates the apparent size of the Moon relative to the Sun. A high ratio often means the more spectacular sights are found near the edge of the track and not on the central line.
- Table 13.2: Annular eclipses. Same data as for totals.
- Table 13.3: Partial eclipses. These are not total or annular anywhere along the track. Times, magnitude of eclipse and areas of eclipse are given.

Tables 13.4 to 13.6 show the lunar equivalents. The data speak for themselves, but in some cases a few extra comments may be helpful.

1998, February 26	*Total.* A Saros on from the brilliant 1980 scenes in Kenya, the westward movement puts this one well out to sea, though it passes through Columbia and Venezuela. These two countries are already hot favourites, and many cruises are planned. The partial phase covers the southern USA, all of Central America and the northern part of South America.
1998, August 22	*Annular.* The track covers Sumatra and North Borneo. The partial phase covers Australia and New Zealand.

1999, February 16	*Annular.* Most of the track extends across the sea, but it does reach Australia. As the Moon has 99% the diameter of the Sun, this is a very short annularity with a very narrow track. The partial phase covers all Australia and New Zealand, with the East Indies and also the southernmost part of Africa and Madagascar.
1999, August 11	*Total.* The British Eclipse. Described separately in Chapter 11.
1999, November 15	*Transit.* Cheating a little, but still the Sun in eclipse although the eclipsing body is so much smaller this time. Transits of Mercury are rare and because of the extreme orbital inclination, November events occur in a ratio of 7 to 3 more frequently than in May. The last one was in 1993, and as on that occasion, Australia is a good place to view it.
2000, February 5	*Partial.* Almost 60% of the Sun is hidden, but as the eclipse is confined to the far Antarctic it is unlikely to be seen by many observers apart from penguins.
2000, July 1	*Partial.* Another Antarctic eclipse. The area does touch the southern tip of South America, including Tierra del Fuego, but here the Sun will be very low in the sky; less than 50% of the Sun is obscured.
2000, July 31	*Partial.* Alaska and parts of northern Canada and Greenland, with northern Asia including Novaya Zemlya are included, making this a good chance to see a "midnight sun" eclipse!
2000, December 25	*Partial.* A Christmas eclipse, which will no doubt be widely seen because it covers virtually the whole of the USA. It does not touch Europe.
2001, June 21	*Total.* 18 years after the Java eclipse, this is another accessible one as the track crosses Africa and Madagascar. It is also "long" with totality lasting nearly 5 minutes at the most favourable sites.
2001, December 14	*Annular.* Virtually the whole track is on the sea; it just reaches Central America, but from here the Sun will be very low.
2002, June 10	*Annular.* Another brief annularity as the Moon has 99% the diameter of the Sun; the track is very narrow and some places provide a good chance to look for some of the phenomena of totality. Unlike the Saros before in 1984 in USA, the best sites are far out to sea. Skilled navigation will be needed for prospective viewers.
2002, December 4	*Total.* South Africa will be the best land site. The partial phase covers a wide area, from Africa down to Antarctica.
2003, May 7	*Transit.* Mercury transits again. With a mid-point in the early morning in the UK (just before 08.00 hrs) a good place to be is further east.
2003, May 31	*Annular.* The last British annular for some time; it just touches North Scotland. A good observing site (clouds permitting) will be Iceland, as from Reykjavík the eclipse is almost exactly central.
2003, November 23	*Total.* Antarctica again, but the partial phase extends over Australia and New Zealand.
2004, April 19	*Partial.* Reasonable views should be had from South Africa.
2004, June 8	*Transit.* Transits of Venus are extremely rare, the last one being in 1882. In June, it is not too bad from the UK with a mid-point at breakfast time, about half-past eight. Better to go a little further east.

2004, October 14	*Partial.* A large partial of over 90%. Japan is in the zone.
2005, April 8	*Annular/Total.* To see totality, lasting for a mere 42 seconds, involves a sea journey. Over most of the track the eclipse is annular.
2005, October 3	*Annular.* Accessible; the track crosses Spain, and right across Africa. The partial phase covers almost all Africa and part of Europe. From London the obscuration amounts to 56% and rather less in Scotland.
2006, March 29	*Total.* An excellent eclipse, a Saros on from the Sumatra event in 1988, fresh in our memory. The track stretches from the Atlantic across Africa, the Mediterranean, Turkey and into Russia. Africa, parts of India and Europe see the partial phase. From London the obscuration is 17%, though only 10% from Edinburgh.
2006, September 22	*Annular.* The track begins in the northernmost part of South America, but then crosses the Atlantic without touching any major land mass. The partial phase can be seen from most of South America and western Africa, including South Africa.
2006, November 8	*Transit.* The best place to see this Mercury transit is the Antipodes and Far East, as the mid-point is after sunset in the UK (just before 22 hrs).
2007, March 19	*Partial.* Much of North America is covered and Japan is at the edge of the eclipse zone.
2007, September 11	*Partial.* Of no particular note although nearly 75% obscuration. It can be seen from much of South America and extends into the Antarctic.
2008, February 7	*Annular.* The Antarctic again (by now the penguins must be getting used to eclipses). New Zealand, Tasmania and southernmost parts of Australia see the partial phase.
2008, August 1	*Total.* A Saros on from the one causing so much bother in Finland in 1990 with its low altitude and cloud, this one is so much better. The track extends from North Canada through Greenland, Novaya Zemlya, Siberia and into China. The partial phase can be seen from much of Europe and the obscuration is 24% from Edinburgh, but only 12% from London.
2009, January 26	*Annular.* The track crosses the Indian Ocean, just reaching the East Indies, but from there the Sun will be very low. The partial phase extends from Central Africa down to Antarctica.
2009, July 22	*Total.* The "Big One" completes its big tour around the Earth before going on to repeat the conditions of 1973 and 1991. Still very long at over $6\frac{1}{2}$ minutes at maximum phase, the track crosses East China and the Pacific, passing not far south of Japan. The partial zone extends over much of China, and down to the East Indies, brushing the tip of Queensland in Australia.
2010, January 15	*Annular.* A nice long one over 11 minutes at maximum with a track over 300 km wide. The track begins in Africa and crosses the Indian Ocean, ending not far from the coast of Asia. As so often happens, the best site lies in mid-ocean. The partial phase covers large areas of Africa and the Far East.

2010, July 11	*Total.* The track lies almost entirely in the Pacific – which is a pity, as this is a long totality of over 5 minutes. It repeats the similar almost impossible conditions in 1992 in the Southern Atlantic. The track does reach the southernmost tip of South America, but from here the Sun will be very low, so that a sea trip is again recommended.
2011, January 4	*Partial.* Europe, Russia and northern Africa will see this one. From London the Sun is 67% eclipsed, but is very low down close to sunrise.
2011, June 1	*Partial.* The Arctic is favoured; Iceland is on the edge of the zone, but the British Isles are just outside.
2011, July 1	*Partial.* A small partial, in duration, obscuration of less than 10%, and geographical area covered. It does not repeat the 1993 conditions as it is the first of a new Saros, number 156, in a series of 69, a rare occasion in a lifetime. As it is restricted to the southernmost Antarctic Ocean it is unlikely to be widely observed.
2011, November 25	*Partial.* Antarctica yet again – almost the entire continent. New Zealand and Tasmania are on the edge of the zone.
2012, May 20	*Annular.* On more familiar territory, this repeats the 1976 conditions in Santorini and 1994 in Arizona but with shorter duration and coverage (under 6 minutes and 95% obscuration). The track begins in China, crosses south Japan and the Pacific, and touches the United States. The Sun's altitude is reasonable from the Japanese area, but very low on the American coast.
2012, June 6	*Transit.* Make the most of this rare second Venus transit in our period as the next will not happen for another century, in 2117. A total non-event from the UK with a mid-transit at half-past one in the morning; the Far East and Australia will be popular.
2012, November 13	*Total.* Repeating the excellent 1976 conditions in Zanzibar and Chile in 1994, the track begins in the northern part of Australia. The track then passes north of New Zealand and extends across the Pacific. Few land masses see the partial phase, apart from the coast of Antarctica.
2013, May 10	*Annular.* Again mainly oceanic, though the track does begin in the northernmost part of Australia. The partial phase covers Australia but only brushes New Zealand.
2013, November 3	*Annular/Total.* This one differs from the 1995 Indian eclipse which was total along the whole track, and is quite a lot shorter in duration. Totality will be seen just off the coast of Liberia and the Ivory Coast. The annular track crosses Africa, and the partial phase covers most of the continent, as well as Spain and the Mediterranean area, but does not extend as far as the British Isles.
2014, April 29	*Annular.* This is a very brief annularity, lasting for less than a second, since the Moon's diameter is 99% that of the Sun. From the region of greatest eclipse – in Antarctica again – the Sun is less than a degree above the horizon. Conditions were even worse in 1996 when the eclipse was only partial in the same part of the world.
2014, October 23	*Partial.* This Saros in 1996 caused much excitement in the UK being the first good partial for a long time. Slightly more than 80% of the Sun is hidden this time in Western America, including Alaska.

2015, March 20	*Total.* The "cold one" repeats the 1979 Montana and 1997 Mongolian conditions, and is an Arctic eclipse, covering Spitzbergen. Most of the track lies in the Arctic Ocean, covering the Faroes but narrowly missing Iceland. The Shetland Islands are not far from the main track and Britain will see a large partial: 85% from London, 93% from Edinburgh.
2015, September 13	*Partial.* Antarctic yet again. Southern Africa lies inside the zone but Australia does not.
2016, March 9	*Total.* The first repeat in our chosen period, duplicating 1980 in Kenya and Venezuela in 1998. The best site will be in the East Indies; the track crosses Borneo and Celebes. The partial zone extends to Australia but not to New Zealand.
2016, May 9	*Transit.* The fourth Mercury transit in our period is almost ideal from the UK with a mid-point in the afternoon, around 15 hrs.
2016, September 1	*Annular.* Accessible, as the track crosses central Africa and Madagascar. All of Africa lies in the partial zone, which extends as far as Arabia.
2017, February 26	*Annular.* Annularity lasts for less than a minute, as the Moon's diameter is 99% that of the Sun. The track begins in the southern-most part of South America and extends across the African coast, but – as usual – the best views should be obtained from mid-ocean.
2017, August 21	*Total.* This should be one of the best-observed eclipses in modern times, as the track crosses the densely populated areas of the United States. The partial phase covers all of North America and extends well into the South American continent. Britain lies at the very edge of the eclipse zone, but Iceland is included.
2018, February 15	*Partial.* Antarctica and southern America only. The partial zone does not reach New Zealand or Australia.
2018, July 13	*Partial.* Like the one in February it is only a small eclipse. The Antarctic Ocean is covered and Tasmania as bits of southernmost Australia are just within the zone.
2018, August 11	*Partial.* A month later and the tables are turned and the Arctic sees this one. Greenland, Iceland, Spitzbergen, Norway, Sweden, Finland and Siberia are in the track, and a very small partial will be seen from the extreme north of Scotland.
2019, January 6	*Partial.* A reasonable view should be had from Japan and parts of China. The zone does not extend as far as North America.
2019, July 2	*Total.* Nowhere near as favourable as 1983 or 2001, as it is almost entirely over the Pacific, though the end of the track does cross South America towards dusk. The partial phase does extend as far as Central America.
2019, November 11	*Transit.* The fifth Mercury transit is not very favourable from the UK. It occurs in the late afternoon, and with a mid-point after 15 hrs, close to sunset, USA is a much better place to be.
2019, December 26	*Annular.* The best view may be expected from Sumatra or Borneo. The track also crosses the southern tip of India, brushing Sri Lanka. A large partial will be seen from Central Australia in the South to the whole of Asia.

| 2020, June 21 | *Annular.* This is the third in the sequence starting in 1984 which caused so much excitement in the USA, being the first well covered by TV, with glimpses of the inner corona and prominences, the second being in 2002. The Moon has 99% the diameter of the Sun. The track crosses Africa, North India and China. The partial zone covers a wide area from the Mediterranean to Japan and the East Indies. |
| 2020, December 14 | *Total.* A nice way to end this review with over two minutes of totality, the track extending from the Pacific across north Chile and Argentina, into the Atlantic; it does not reach Europe. The partial phase covers almost all of South America. |

The five transits of Mercury are a very special form of solar eclipse, an event not to be missed. Even more exciting because of their extreme rarity are the two transits of Venus, which occur in pairs over a century apart.

Make a special effort to see these fascinating events. Solar projection is by far the best way of following and photographing the progress. Venus transits are noted for the famous "black-drop" effect whereby the planetary limb appears to bleed-in, making it look like a pear or tear drop. Some of this is due to optical defects in the optics and the eye, some due to atmospheric turbulence. These events will be a good opportunity to study all optical and eye illusions with the latest electronic imaging and optical design.

Let us all hope for clear skies!

Table 13.1. Total eclipses

Date	Max eclipse Lat.	Long.	Ut h-m-s	Duration min-sec	Altitude degrees	Track width miles	km	Ratio	Major areas
1998 Feb 26	04° 43'N	82° 43'W	17-28-22	4-09.7	76	94	151	1.04	Pacific, Columbia, Venezuela, Atlantic
1999 Aug 11	45° 04'N	24° 18'E	11-03-05	2-22.8	59	70	112	1.03	Atlantic, W.England, Alderney, Europe, India
2001 Jun 21	11° 16'S	02° 45'E	12-03-42	4-56.6	55	124	200	1.05	Atlantic, Southern Africa, Madagascar
2002 Dec 04	39° 28'S	59° 34'E	07-31-08	2-03.7	72	54	87	1.02	Southern Africa, Indian Ocean, Australia
2003 Nov 23	72° 41'S	88° 29'E	22-49-15	1-57.2	15	308	496	1.04	Antarctica only
2005 Apr 08	10° 34'S	118°58'W	20-35-41	0-42.0	70	17	27	1.01	Pacific, Central America
2006 Mar 29	23° 09'N	16° 46'E	10-11-15	4-06.7	67	114	184	1.05	Atlantic, Ghana, Niger, Libya, Turkey, Russia
2008 Aug 01	65° 38'N	72° 16'E	10-20-59	2-27.2	34	147	237	1.04	N.Canada, Greenland, ending in N.China
2009 Jul 22	24° 12'N	144°08'E	02-35-12	6-38.9	86	161	259	1.08	E.China, Pacific
2010 Jul 11	19° 47'S	121°51'W	19-33-25	5-20.2	47	161	259	1.06	Pacific, extremes of Chile & Argentina
2012 Nov 13	39° 58'S	161°18'W	22-11-39	4-02.2	68	111	179	1.05	Queensland, Pacific
2013 Nov 03	03° 29'N	11° 40'W	12-46-20	1-39.6	71	36	58	1.02	Atlantic, Central Africa
2015 Mar 20	64° 25'N	08° 34'W	09-45-30	2-46.9	19	287	462	1.04	N.Atlantic, Arctic, Spitsbergen
2016 Mar 09	10° 07'N	148°50'E	01-57-02	4-09.5	75	96	155	1.04	East Indies, Pacific
2017 Aug 21	36° 58'N	87° 38'W	18-25-21	2-40.1	64	71	115	1.03	Pacific, Central USA, Atlantic
2019 Jul 02	17° 24'S	108°57'W	19-22-46	4-32.7	50	125	201	1.05	Pacific, Chile, Argentina
2020 Dec 14	40° 21'S	67° 54'W	16-13-15	2-09.7	73	56	90	1.03	Pacific, Chile, Argentina, Atlantic

Table 13.2. Annular eclipses

Date	Max eclipse		Ut	Duration	Altitude	Track width		Ratio	Major areas
	Lat.	Long.	h-m-s	min-sec	degrees	miles	km		
1998 Aug 22	02°59'S	145°23'W	02-06-06	3-14	75	61	99	0.973	East Indies, Pacific
1999 Feb 16	39°50'S	93°54'E	06-33-34	0-40	62	18	29	0.993	Indian Ocean, Australia
2001 Dec 14	00°37'N	130°41'W	20-51-53	3-53	66	78	126	0.968	Pacific, Costa Rica, Honduras
2002 Jun 10	34°33'N	178°37'W	23-44-15	0-23	78	8	13	0.996	Pacific
2003 May 31	66°49'N	24°00'W	04-08-16	3-37	3	2790	4498	0.938	Arctic, Greenland, Iceland, N.Scotland
2005 Oct 03	12°52'N	28°45'E	10-31-37	4-31	71	101	162	0.958	Spain, Libya, Sudan, Kenya
2006 Sep 22	20°40'S	09°03'W	11-40-04	7-09	66	162	261	0.935	Surinam, Guinea, Atlantic
2008 Feb 07	67°35'S	150°27'W	03-54-57	2-12	16	276	445	0.965	S.Pacific, Antarctica
2009 Jan 26	34°05'S	70°16'E	07-58-30	7-54	73	174	280	0.928	S.Atlantic, Indian Ocean, East Indies
2010 Jan 15	01°37'N	69°20'E	07-06-23	11-8	66	207	333	0.919	Uganda, Kenya, Indian Ocean, China
2012 May 20	49°05'N	176°19'E	23-52-38	5-46	61	147	237	0.944	S.China, Japan, Pacific, USA
2013 May 10	02°12'N	175°31'E	00-25-05	6-04	74	107	173	0.954	Australia, Solomon Islands, Pacific
2014 Apr 29	70°42'S	131°10'E	06-03-16	0-00	1	0	0	0.986	Antarctica
2016 Sep 01	10°41'S	37°48'E	09-06-43	3-06	71	62	100	0.974	Atlantic, Africa, Madagascar, Indian Ocean
2017 Feb 26	34°42'S	31°08'W	14-53-13	0-44	63	19	31	0.992	Chile, Argentina, Atlantic, Central Africa
2019 Dec 26	01°00'N	102°18'E	05-17-28	3-39	66	73	118	0.970	Arabia, Sri Lanka, Sumatra, Borneo
2020 Jun 21	30°31'N	79°43'E	06-39-52	0-38	83	13	21	0.994	Africa, India, China, Taiwan

Table 13.3. Partial eclipses

Date	Max eclipse Ut h-m-s	Magnitude %	Major areas
2000 Feb 05	12-49-20	58	Antarctic
2000 Jul 01	19-32-29	48	Antarctic Ocean, tip of South America
2000 Jul 31	02-13-04	60	N. Asia, Greenland, North Canada
2000 Dec 25	17-34-49	72	USA, Central America, Canada, Greenland
2004 Apr 19	13-33-58	74	Antarctica, S. Atlantic, Southern Africa
2004 Oct 14	02-59-14	93	N.E. Asia, Japan, Alaska
2007 Mar 19	02-31-49	87	E. Asia, Japan, Alaska
2007 Sep 11	12-31-13	75	South America, Antarctica
2011 Jan 04	08-50-26	86	Europe, N. Africa, W. Asia
2011 Jun 01	21-16-03	60	Japan, N.E. Asia, N. Canada, Greenland
2011 Jul 01	08-38-16	10	Antarctic Ocean
2011 Nov 25	06-20-08	90	Antarctica, New Zealand
2014 Oct 23	21-44-20	81	North America, N.E. Pacific
2015 Sep 13	06-53-59	79	Southern Africa, Madagascar, Antarctica
2018 Feb 15	20-51-11	60	Antarctica, Chile, Argentina
2018 Jul 13	03-00-55	34	Victoria, Tasmania, Antarctic Ocean
2018 Aug 11	09-46-07	74	Greenland, Scandinavia, N. Asia
2019 Jan 06	01-41-14	71	N.E. Asia, Japan, N. Pacific

These eclipses are not total or annular anywhere on Earth.

Table 13.4. Total lunar eclipses

Date	Start Ut h-m	Middle	End	Duration h-m	Magnitude %	Major areas for mid eclipse
2000 Jan 21	04-04	04-44	05-22	1-18	133	N. & S. America, UK, W. Europe
2000 Jul 16	13-02	13-56	14-49	1-47	177	Japan, whole of W. Oceania
2001 Jan 09	19-50	20-21	20-52	1-02	119	Europe, Africa, Asia
2003 May 16	03-14	03-40	04-07	0-53	113	E. USA, S. America
2003 Nov 09	01-07	01-19	01-31	0-24	102	Eastern N. & S. America, UK, Europe, W. Africa
2004 May 04	19-52	20-30	21-08	1-16	131	Central Africa & Asia, Madagascar
2004 Oct 28	02-23	03-04	03-45	1-22	131	N. & S. America, UK, Western Europe & Africa
2007 Mar 03	22-44	23-21	23-58	1-14	124	UK, Africa, W. Asia
2007 Aug 28	09-52	10-37	11-23	1-31	148	Central Oceania, New Zealand, Alaska
2008 Feb 21	03-00	03-26	03-51	0-51	111	E. USA, S. America, UK, W. Europe & Africa
2010 Dec 21	07-40	08-17	08-54	1-14	126	N. America, Central Oceania
2011 Jun 15	19-22	20-13	21-03	1-42	171	Central Africa & Asia
2011 Dec 10	14-06	14-32	14-58	0-52	111	E. Asia, East Indies, Australia
2014 Apr 15	07-06	07-46	08-25	1-21	130	USA, Canada, Western S. America, Central Oceania
2014 Oct 08	10-24	10-54	11-24	1-00	117	Western N. America, Central Oceania
2015 Apr 04	11-56	12-00	12-05	0-09	101	E. Australia, New Zealand, Central Oceania
2015 Sep 28	02-11	02-47	03-24	1-13	128	E. USA, S. America, UK, Western Europe & Africa
2018 Jan 31	12-51	13-30	14-08	1-17	132	Eastern Asia, East Indies, Australia, New Zealand
2018 Jul 27	19-30	20-22	21-14	1-54	161	Central Africa & Asia, Madagascar
2019 Jan 21	04-41	05-12	05-44	1-03	120	N. & S. America, UK, Iceland

All these eclipses are total, with the Moon fully in the umbra. The first contact with the penumbra averages 2 hours before the start of totality, and last contact with the penumbra another two hours after the end of totality. The total time taken for all the phenomena to be completed is often well over 5 hours.

Table 13.5. Partial lunar eclipses

Date	Start Ut h-m	Middle	End	Duration h-m	Magnitude %	Major areas for mid eclipse
1999 Jul 28	10-22	11-34	12-45	2-22	40	Oceania
2001 Jul 05	13-35	14-55	16-15	2-40	50	Australia, Eastern Asia, East Indies, New Zealand
2005 Oct 17	11-34	12-03	12-32	0-58	7	New Zealand, Far East, Australia & E. Indies
2006 Sep 07	18-05	18-51	19-38	1-33	19	Central Asia, E. Africa, W. Australia
2008 Aug 16	19-35	21-10	22-44	3-09	81	Africa, India, Madagascar
2009 Dec 31	18-52	19-23	19-54	1-04	8	UK, Europe, Africa, Asia
2010 Jun 26	10-16	11-38	13-00	2-44	54	E. Australia, New Zealand, Central Oceania
2012 Jun 04	09-59	11-03	12-07	2-08	38	E. Australia, New Zealand, Central Oceania
2013 Apr 25	19-52	20-07	20-22	0-30	2	Central Asia & Africa, Madagascar
2017 Aug 07	17-22	18-20	19-18	1-56	25	Central Asia, E. Africa, W. Australia
2019 Jul 16	20-01	21-31	23-00	2-59	66	Africa, Middle East, W. India

All these eclipses are partial with the Moon's disk never fully into the umbra, but completely immersed in the penumbra. The magnitude is the percentage immersion in the umbra, which can be very low as in 2013, to nearly total as in 2008. The first contact with the penumbra averages $1\frac{1}{2}$ hours before the first contact with the umbra, the start of the partial eclipse, and last contact with the penumbra another $1\frac{1}{2}$ hours after the last contact with the umbra, the end of the partial eclipse proper. The total time taken for all the phenomena to be completed is similar to a total eclipse, often well over 5 hours.

Table 13.6. Penumbral lunar eclipses

Date	Start Ut h-m	Middle	End	Duration h-m	Magnitude %	Major areas for mid eclipse
1998 Mar 13	02-14	04-20	06-26	4-12	73	N. & S. America, UK, W. Africa
1998 Aug 08	01-32	02-25	03-17	1-46	15	E. USA, S. America, UK, Europe, Africa
1998 Sep 06	09-14	11-10	13-06	3-52	84	Australia, New Zealand, Oceania, Alaska
1999 Jan 31	14-04	16-18	18-30	4-26	103	Asia, E. Indies, Australia
2001 Dec 30	08-25	10-29	12-33	4-08	92	N. America, Japan, New Zealand
2002 May 26	10-13	12-03	13-54	3-41	71	Japan, Australia, New Zealand
2002 Jun 24	20-18	21-27	22-35	2-16	23	Europe, India, Africa
2002 Nov 20	23-32	01-47	04-01	4-29	89	UK, most of USA, S. America, Europe & Africa
2005 Apr 24	07-50	09-55	12-00	4-10	89	Western N. America, New Zealand, Oceania
2006 Mar 14	21-22	23-47	02-14	4-52	106	UK, Europe, Africa
2009 Feb 09	12-37	14-38	16-39	4-03	92	Asia, Australia, New Zealand
2009 Jul 07	08-33	09-39	10-44	2-11	18	Western N. & S. America, Oceania
2009 Aug 06	23-01	00-39	02-17	3-16	43	UK, Europe, Africa, S. America
2012 Nov 28	12-12	14-33	16-53	4-41	94	Asia, East Indies, Australia
2013 May 25	03-44	04-10	04-36	0-53	4	USA, S. America, W. Africa
2013 Oct 18	21-48	23-50	01-52	4-03	79	UK, Europe, Africa, Brazil
2016 Mar 23	09-37	11-47	13-57	4-20	80	Alaska, Japan, E. Australia, New Zealand
2016 Aug 18	09-26	09-42	09-59	0-34	2	N. America, most of S. America & Oceania
2016 Sep 16	16-53	18-54	20-56	4-03	93	Asia, most of Africa & Australia
2017 Feb 11	22-32	00-44	02-55	4-23	101	UK, Europe, Africa, Brazil
2020 Jan 10	17-05	19-10	21-14	4-09	92	UK, Europe, Asia, most of Africa
2020 Jun 05	17-43	19-25	21-06	3-23	59	S. Asia, E. Africa, W. Australia
2020 Jul 05	03-04	04-30	05-55	2-51	38	S. America, most of N. America, Antarctica
2020 Nov 30	07-30	09-43	11-56	4-26	85	N. America, Oceania

The Moon's disk never enters the umbra, and sometimes is not even completely in the penumbra, indicated by a magnitude of less than 100%. They may be extremely difficult to detect when the atmosphere around the Earth's horizon is very clear. If the penumbral eclipses of the Sun by the Earth in 2013 and 2016 are 'bright', they will be a supreme test of observational skill.

Chapter 14

Eclipses Elsewhere

We Earth-dwellers are fortunate in that we have a Moon just – and only just – large enough to cover the Sun. If the Moon loomed much smaller in our sky, there would be no total eclipses – only annulars and partials – and the very experience of the corona might have remained unknown until the start of the Space Age. If the Moon appeared much larger, there would certainly be protracted total eclipses, but they would be shorn of their beauty.

Nobody has yet observed an eclipse of the Sun by the Earth, which, obviously, can happen only at the time of our full moon, but no doubt will be observed as soon as we establish a permanent Lunar Base. Meanwhile, what about eclipses from other planets?

Mercury and Venus have no satellites. Mars, as we have seen, has two – Phobos and Deimos – but both are very small, and are irregular in shape. From Mars, the Sun has a mean apparent diameter of 21 minutes of arc; as seen from the Martian surface Phobos is never more than 12.3 minutes of arc across, and Deimos only about 2 minutes of arc. Both will transmit the Sun frequently, since their orbital inclinations are negligible. Phobos will do so 1300 times in every Martian year, taking 19 seconds to cross the Sun's disk; there will be 130 transits of Deimos, each taking 1 minute 48 seconds. In fact, satellite transits will be so common that the Martian colonists of the coming century are unlikely to take more than a very casual interest in them, if that!

For the giant planets the situation is different, and needless to say, there can never be any question of landing there, so that eclipses as seen from their surfaces (or, rather, their upper clouds) are of theoretical interest only. In fact the satellite families could cause transits often. Details are given in the following table:

JUPITER	Apparent diameter of Sun:	6'	09"
	Amalthea	7	24
	Io	35	40
	Europa	17	30
	Ganymede	13	06
	Callisto	9	30
SATURN	Apparent diameter of Sun:	3'	22"
	Mimas	10	54
	Enoeladus	10	36
	Tethys	17	36
	Dione	12	24
	Rhea	10	42
	Titan	17	10
URANUS	Apparent diameter of Sun:	1'	41"
	Miranda	17	54
	Ariel	20	54
	Umbriel	14	12
	Titania	15	00
	Oberon	9	48
NEPTUNE	Apparent diameter of Sun:	1'	04"
	Triton	26	13

All these could produce total eclipses; the remaining planetary satellites could not. But in no case is there a satellite diameter which is virtually the same as that of the Sun. Amalthea, in Jupiter's system, comes nearest.

One day there may be expeditions to some of these satellites, with the possibility of seeing solar eclipses caused by other satellites; but anything of the kind is a long way ahead yet. When – or if – we will reach these remote worlds remains to be seen.

So far as we are concerned, it has often been asked whether the similarity of the apparent diameters of the Sun and the Moon is sheer coincidence. The answer can only be "yes" – but it is a fortunate coincidence for us.

In this book we have discussed eclipses in all their aspects, and we have – we hope – been able to give useful advice to those who intend to travel eclipse tracks in the coming years. There is always something new to see; and there is nothing in the whole of Nature to rival that magic moment when the last segment of the Sun's brilliant disk disappears, and the corona and prominences flash into view.

Appendix A

Cassette Loads

Sometimes the disadvantages of "rolling your own" film from bulk are outweighed by the practical advantages, for example when expense is a major consideration.

Use the table below to calculate how many cassettes can be loaded from bulk film, or the converse after exposure to calculate the length of spliced film when this economical processing option is possible. (30 metres is the metric equivalent of 100 ft.)

Obviously fractions of a cassette are illogical, but Table A.1 does say how usable the reel-end should be, and conversely how easy it will be to squeeze in an extra load by shortening the leader if the camera does not need as much as usual.

There are many types of reloading or winding apparatus available, and it is sad to say that many will scratch film or damage it in some nasty way. Only buy from a reputable shop or where you can give it a good inspection to see if the technique is workable back at the ranch.

The better machines allow a direct reel-to-reel operation, and whilst these usually mean working in total

Table A.1. Cassette loads

| Exposures | Length | | Per 80 | Per 100 | Per 150 |
	feet	cm	ft	ft	ft
12	2.25	68.6	35.5	44.4	66.6
20	3.25	99.1	24.6	30.7	46.1
24	3.75	114.3	21.3	26.6	40
36	5.25	160	15.2	19	28.5
50	7	213.4	11.4	14.3	21.4

darkness, they do offer much more confidence and freedom from scratching as film only goes through the felt lips for the first time in the camera. This can be a major consideration in coronal studies where low light exposures are intended. The peace of mind from knowing that pristine bulk film with guaranteed characteristics is loaded directly into cassette at home can far outweigh other considerations for unique experiments. Other practical advantages include:

1. Plastic cassettes carried in the pocket will not activate metal detectors at security checks. This may be the only practical way of ensuring that very high speed and special films are not X-rayed in luggage.

2. Take full advantage of this metal detector blindness by transferring commercial film in metal cassettes into these useful devices. However, never try this dodge on important film without practice and practice again, until every reel can be transferred perfectly, without scratching and pressure marks. Practice is necessary since not all spools from a metal cassette will fit properly into the plastic ones, and light security might be compromised. There may be no alternative to winding off the original spool then back onto the new one. Remember, the old leader is light exposed and could end up carrying the vital last exposure – think about it, now, not on eclipse day!

3. Squeezing in as many frames as possible to cope with motor drives' insatiable thirst for sequence shots. Plastics cassettes only hold the 'official' top limit of 36 exposures and often not that many when confronted with motor drives, whereas metal ones will usually take up to 45 with some care in operating the winding mechanism. Film thickness is the deciding factor.

4. Alternatively, go for economy and load just the number needed (with a safety margin, of course) for special experiments.

5. In many countries some of the more useful professional films are only readily available as bulk lengths and there is no alternative to winding into cassettes at home. Such films include ciné and tungsten copying films needed for duplicating valuable originals or CCD/video/computer monitor screens.

6. Some films are coated on Estar base which is much thinner than the normal acetate, allowing up to

50% more film to be crammed in. Tech Pan and lith are two obvious types with special spectral characteristics quite useful in monochrome experiments, and daily routine monitoring. 50 exposure cassettes are easy with these films.

Appendix B

Cloud Types

Knowing your enemy is always a good adage, and it can make all the difference in last minute plans. Any cloud will destroy much of the critical science, whilst actually enhancing the visual spectacle to the extreme annoyance of the scientists present.

Meteorologists use a classification system based on "vertical extent" and "height and consistency". This means little to the average person, but we need to use their terms to understand the cloud types which are segregated into three levels: 20,000 ft and over, 8000–20,000 ft, and up to 8000 ft. Local weather forecasts will, or at least should, give information on the cloud types according to these categories, and the longer-range the forecast the better to take evasive action.

High Clouds 6000 metres (20,000 ft) and over

1. *Cirrus*. Entirely ice crystals and seen as thin white strands to long sheaves.
2. *Cirrostratus*. Also ice crystals, seen as flattish, layered, milky-coloured clouds. Sometimes seen with a pearly halo effect caused by refraction of light through the ice crystals.
3. *Cirrocumulus*. Caused by convection currents, these are very high, like fluffy pieces of cotton wool.

All clouds in this category can be tolerated to some extent in visual work and most photography. The inner corona and prominences will show through 1 and 2 quite well (Fig. B.1) and the spectacle of a coloured halo round the eclipsed sun was one of the major features of the 1979 Montana eclipse. Bracket well and start with 1–2 stops for absorption.

Fig.B.1. A partial eclipse observed from Alderney in 1996. Slightly cloudy conditions still allow the eclipse to be seen. *(Michael Maunder)*

Medium Clouds 2500–6000 metres (8000–20,000 ft)

1. *Altocumulus.* These are variable. Basically, like cirrocumulus but larger, darker grey, sometimes towering from a broad base.

2. *Altostratus.* Like cirrostratus but lower. Usually lighter grey, and sometimes the Sun can be seen through the cloud. If so, there may be a corona around the Sun.

Clouds in this category are best avoided. Some Type 2 was present in Montana to complicate the exposures.

Lower Level Bases up to 2500 metres (8000 ft)

1. *Stratus.* Usually quite low and not very thick, seen as a grey amorphous mass, often ragged. When at ground level it is fog.

2. *Stratocumulus.* Like altocumulus, but individual shapes can be picked out. Flattish on top.

3. *Cumulus.* Caused by convection. Like small, fluffy bits of cotton wool, but can grow into giant cauliflowers causing rain, and can change to...

   Cumulonimbus. Towering clouds with rain at bottom, and sometimes thunderstorms. The top can be fibrous-looking through picking up ice crystals, and the tops of the biggest are often anvil-shaped. From below, the bottom looks black, and from one side, the sides and top look white. Anything in this category is bad news and every effort should be made to go elsewhere.

   Unfortunately, Types 2 and 3 are the most likely clouds in Cornwall for the 1999 eclipse, as hot and humid air from the sea rises over cliff faces and higher ground, particularly as the air cools towards totality. The effect has been likened to the "English Monsoon", when thunder is common during the month. A stable high pressure system and no wind is the best cure, although fog then becomes an alternative hazard.

N.B. There is another sub-type in the medium level called *nimbostratus*, which is a grey amorphous mass always associated with rain. Where you see "nimbo" you see rain, and this makes it another weather system to avoid. The medium level is mainly associated with frontal conditions indicating a change in the weather.

There is always an element of luck in eclipse chasing and recent years have seen a complete pattern reversal around eclipse days, with dry areas pouring rain, and cloudy regions clearing as if by magic but only on the day, for the day. It also pays to keep in mind that the region behind a cold front can often give the clearest and finest skies before the warm one moves in. Mobility is the name of the game.

Appendix C
Suggested Total Solar Eclipse Exposures

Focal Ratio	Baily's Beads	Diamond Ring	Prominences	Inner corona	Outer corona	Landscape	Partials ND=5
25 ISO Film							
2	1/4000	1/1000	1/500	1/60	1/8	1/4	1/4000
4	1/1000	1/250	1/125	1/15	1/2	1	1/1000
8	1/250	1/60	1/30	1/4	2	4	1/250
16	1/60	1/15	1/8	1	8	16	1/60
50 ISO Film							
2	1/8000	1/2000	1/1000	1/125	1/15	1/8	1/8000
4	1/2000	1/500	1/250	1/30	1/4	1/2	1/2000
8	1/500	1/125	1/60	1/8	1	2	1/500
16	1/125	1/30	1/15	1/2	4	8	1/125
100 ISO Film							
2	—	1/4000	1/2000	1/250	1/30	1/15	—
4	1/4000	1/1000	1/500	1/60	1/8	1/4	1/4000
8	1/1000	1/250	1/125	1/15	1/2	1	1/1000
16	1/250	1/60	1/30	1/4	2	4	1/250
400 ISO Film							
2	—	—	—	1/1000	1/125	1/60	—
4	—	1/4000	1/2000	1/250	1/30	1/15	—
8	1/4000	1/1000	1/500	1/60	1/8	1/4	1/4000
16	1/1000	1/250	1/125	1/15	1/2	1	1/1000
1000 ISO Film							
2	—	—	—	1/2000	1/250	1/125	—
4	—	—	1/4000	1/500	1/60	1/30	—
8	—	1/2000	1/1000	1/125	1/15	1/8	—
16	1/2000	1/500	1/250	1/30	1/4	1/2	1/2000
3200 ISO Film							
2	—	—	—	—	1/1000	1/500	—
4	—	—	—	1/2000	1/250	1/125	—
8	—	—	1/4000	1/500	1/60	1/30	—
16	—	1/2000	1/1000	1/250	1/15	1/8	—

It is possible to work out eclipse exposures from first principles on the assumption that the prominences are almost as bright as the full moon, the inner corona is 3 stops fainter, and the outer corona another 3 stops down. The table assumes that the prominences are not quite that bright and also that the sky is of excellent clarity and the lens and film characteristics are within specification.

Always bracket and make due allowance for sky clarity and other factors, although it should be painfully obvious from the vast range of options listed that it is extremely difficult not to get something even if way off the original intention.

1. Exposures must be spot-on with reversal films, which dictates more bracketing than ever. Print films have considerably more latitude, and two stops either way of the claimed ISO will rarely be noticed except in the most critical of printing machinery. Over-exposure with print film is less detrimental than under-exposure.

2. Halation from the prominences, inner corona and chromosphere onto the black solar disk is always a problem with the longer exposures necessary for the outer corona. The highest quality optics and the longest practical focal length to fit into the frame size are prime requisites for this sort of work. Medium format equipment is almost obligatory.

3. The same rules apply with Baily's Beads and the Diamond Ring. The table makes some allowance for halation effects and these exposure times are preferable to the normally published lists with shorter focal lengths and fast optical systems. Longer focal length systems, say over a metre and f ratios of $f16$ or slower, particularly on the larger formats do benefit from an increased exposure of about a stop to enhance the phenomena, particularly when using print film or CCD/video. Many of the poor quality images seen on CCD/video can be traced to halation and gross overexposure due to too small an image scale, where gross overload swamps everything.

4. Photographing landscapes at totality is always great fun and well worth the effort. Something will be captured with the exposures used for the outer corona, but experience indicates that a stop extra does some good lightening the foreground 'gloom',

and these figures are used. The bright sky outside the Moon's shadow is never intense enough to burn out completely even if it is grossly over-exposed relative to the main scene. Always use a small stop to gain depth of focus, and a slow film to enhance subtle colours.

5. Shadow bands, if these elusive things appear, will be caught on the same exposures used for the landscape.

6. Working on figures for the partial phases in the weeks before the trip is much too late by far! Partial photography is one thing which can be sorted out any time and must be second nature in order to devote full attention to the more interesting total phenomena.

Neutral Density filters vary enormously, and the basic one of 5 is given but the actual value must be found. Note, the densities vary as a log function, not as stops, so ND4 = 10 times more light through or $3\frac{1}{2}$ stops, and ND6 = 10 times less light, so a slight difference in specifications means an enormous variation in practice.

Appendix D
Field of View of Camera Lenses

Deciding which lens to use is often a simple matter of economics, luggage restrictions and handling.

A 28 mm on a 35 mm camera is popular for capturing the lunar shadow as it moves across the sky, whereas a 1000 mm is usually accepted as the longest practical length to keep prominences within frame. The tables and notes should help in planning the widely different interests.

35 mm Cameras

Focal length (mm)	Width (degrees)	Height (degrees)	Comments
7.5	Fish-eye = total sky cover		For self-portraits!
15	ditto		
17	93	70	Becoming quite common
20	84	62	Special effects easily caught
24	74	53	Best all-round for sky effects
28	65	46	Regarded as the "norm"
35	54	38	Huge range made
50	40	27	The standard lens
85	24	16	Used like a standard
100	20	14	The bottom end of telephoto
135	15	10	The classic telephoto
200	10	7	Much under-rated
300	6.50′	4.35′	Beginning to be useful
longer	see notes		Difficult to keep in frame

Medium Format

Focal length mm	6 × 9 cm degrees	6 × 7 cm degrees	6 × 6 cm degrees	Comments
30	fish-eye = virtual all round cover of all film sizes			
50	89	81	75	Ideal for sky effects
75	77	68	62	75/80 is the standard for 6 × 6
100	53	47	43	The standard for 6 × 7
127	43	38	35	The standard for 6 × 9
150	37	32	29	Useful for timed sequences
250	22	20	19	Similar cover to 135 on 35 mm

Notes

1. The 35 mm camera figures are taken from typical manufacturers' literature, and make some allowance for cropping and viewfinder limitations, Although the full frame is 36 × 24 mm, the average cropping of 5% reduces this to around 34.2 × 22.8 mm, meaning that these "viewfinder" views are more realistic and somewhat smaller than those often quoted in many astrophotography and astronomical sources.

2. Calculation of the 35 mm camera angles is notoriously inaccurate at the shorter focal lengths, and only the manufacturer's literature or practical measurements should be used below about 100 mm. At the longer lengths (f), the angular widths and heights can be calculated from very approximate "easy-to-remember" figures:

    $$°Width = 2000/f \text{ and } °Height = 1300/f$$

    In applying this calculation, bear in mind that the apparent solar disk at 1/2° has become an appreciable fraction of the frame, and at 1000 mm it is almost 1/3 the frame width and 1/2 the frame height, leaving no margin for tracking and drive errors.

3. The medium format table appears to give a better coverage than 35 mm, and this is true because slides and prints suffer much less cropping and viewscreens can be used as a double check. 6 × 6 standard lenses are more variable and the coverage angles best checked against the maker's literature.

4. As with 6 × 6, no figures are given here for larger format configurations, as the variations are enormous, the owners generally have a better idea of what they are about and the manufacturer's literature is more helpful. Wide-angle studies greatly benefit from the higher resolution and image size, and telephotos can tolerate much more tracking and drive errors whilst still keeping large images on screen.

The classic technique of a century ago applied the philosophy to its logical end. A whole plate (a 10 × 8 is the nearest equivalent today) and a focal length of 1 to 3 metres when propped up on rocks was bound to capture the image somewhere as the Earth rotated. Today's 35 mm users spend much time trying to find the image in the first place.

Sun/Moon Transit Times

The tables list an approximate time on an undriven mount for the Sun or Moon to cross the picture frame with the more popular lens and film format combinations.

Use them to plan a timed sequence of multiple exposures, or to place the image somewhere close to the frame centre at maximum phase or totality.

Times are hours and minutes.

35 mm Cameras

Focal length, mm =	28	50	135	300	500
Along 36 mm frame	3-40	2-15	0-55	0-22'30"	0-13'30"
Across 24 mm frame	2-35	1-35	0-37	0-13'00"	0-7'55"

Lubitel and Other Medium Format Cameras

Focal Length, mm =	50	75	80	100	250
Across 6 cm frame	3-15	2-15	2-05	1-55	0-45
Across 7 cm frame	3-45	2-35	2-25	2-15	0-50
Across 9 cm frame	4-55	3-25	3-05	2-55	1-00

The Sun's image takes a little over 2 hours to cross the film frame with a standard lens which is intended to reproduce the naked-eye view of about 45 degrees across. As the Sun appears to traverse 15 degrees an hour, common sense says it should take 3 hours to move across the frame, but this makes no allowance for the peculiarities of lens and camera manufacture. The lens focal length can be out by as much as 10%, which is quite normal when it is designed to focus from infinity to nearby.

The tables list the more popular combinations from practical experience (and are not necessarily consistent with an arithmetic calculation, because of the factors below) and are no more than a starting point when testing anything to be used in practice.

Notes

1. Many 35 mm camera viewfinders crop the image considerably more than 5% on each side to allow a good safety margin, as many commercial printers crop even more in the final print. When checking transit timings, it is always better to use the viewfinder scene's crop with print films, unless and only if the camera has facilities for checking the actual image with a view screen *and* the film can be printed professionally as a whole frame.

2. Interpretation of what is regarded as a "reasonable" clearance at the frame edge will vary between people, and there are no set rules here. The tables have been made with the whole of the solar image clearly into the film frame as it would be mounted and projected as a slide.

3. There are few places in the world, or days in the year, where the Sun's apparent path across the sky is a true straight line. The transit times are based on the measurements of a straight line crossing as happens close to noon, and with the camera set at the appropriate rising or setting angle.

4. The Sun's movement is usually curved, and the true transit times will be somewhat longer, up to 1.4 times in the extreme case of a corner-to-corner motion across a square.

5. With all the variables, the only sensible way forward is to make test exposures with the Sun transiting at the expected angles and camera orientation.

6. Because backdrops are not always known until nearer the day, a final dry run a day or so before at site is always the best advice, although not always feasible. A dry run should be entered into the plans for the day or so before, as small changes in latitude and longitude will often be much smaller errors than calculations back home. This is particularly important when changing hemispheres, as many a photographer has been caught out by the "flipped" images due to the reversed apparent motion of the Sun. TLR cameras with their laterally reversed viewfinders, and larger format viewscreens need special care when at an unfamiliar hemisphere or site.

7. These times are most useful for standard and wide-angle lenses. For telephotos the times are more dubious and a guidance only, as the Sun's apparent diameter comes into the equation and at extreme focal lengths it is the critical feature which can only be selected by experiment.

Appendix F

Check Lists of Items to take on Eclipses

Any check list is personal bias, but what follows is based on experience.

Copy the lists, select those items which make sense and then renumber into a personal order of priorities. Always add extra items which seem important when they occur, never leave writing them down to a later date.

Pay particular attention to battery power. So many listed items take batteries that much of the baggage allowance will be taken up with them. Chew over the merits of primary batteries which can be bought cheaply and left behind, against rechargeables which can be "topped up" just before the eclipse from the local mains, or solar charger.

For the Eclipse

Item	Check ✓
Vital Items	
Celebration provisions (see text)	
Solar filters for naked-eye views of the partial phases	
Binoculars and/or telescope for totality	
Solar filters to match for the partial phases	
Cameras	
■ "fun"	
■ video	
■ still (conventional)	
■ ciné	
Standard lens for each type of camera	
Video tapes	
Film for each camera	
■ 100 ISO (preferably slower)	
■ high speed, e.g. 400 ISO	
■ Specialist types	
Batteries	
Charger if appropriate	
Sunglasses	
Proper clothing for eclipse site (see text)	
Seating	
Refreshments (see text)	
A natural break!	
"Essentials" for the Serious	
Sturdy tripod for each type of camera	
Sturdy tripod head for each camera	
Spare batteries for tripod equatorial drives	
Special lenses for each type of camera	
■ Telephoto	
■ Wide-angle	
Special mounts for lenses to cameras	
Lens cap for each type of lens	
Lens hood for each type of lens	
Solar filter for each lens	
■ Is it firmly attached in a full gale?	
■ Can it be removed and replaced quickly?	
White cloth to put over equipment	
Spare batteries for each camera body	
Spare batteries for each motor drive	

Item	Check ✓
Cable release for each camera	
Instructions (copies for each and every piece of equipment)	
Battery-operated tape player and headphones	
Detailed written observation check list regardless	
Battery-operated tape recorder for results	
▪ Spare tapes for recorders	
▪ Spare batteries for recorders	
Note book and pencil regardless	
Preferably all on a lanyard round neck	
Small torch, with lanyard to hang round neck	
Spare batteries for torch	
Sensible Precautions	
Puffer brush for cleaning lenses	
Lens cleaning tissues or cloth	
Cotton buds	
Reel of adhesive tape	
▪ Clear adhesive	
▪ Masking	
Blu Tack	
Tweezers	
Jeweller's screwdriver set	
Set of spanners for tripod and other gadgets taken	

For the Rest of the Trip

Item	Check ✓
Medical necessities	
More film	
More video tapes	
Flash gun – do not take on site (see text)	
And yet more batteries!	

Index